美少女CG網站
Anime CG Web Banner Collection

伊東秋

Aki Ito

前言

基本概念

這是一本以收集網路標題為主的作品，原則上是集自己創作的CG及網路上畫家的網址而成的，讀者在閱讀這些網路標題會從中產生種種的閱讀樂趣，而本書就是想像這些樂趣所集聚而成的。把它放在手中，迅速地翻閱，不就和上網路看索引一樣，有著不同的樂趣嗎？一般介紹網路資訊的書籍，在一推出時，就成了過時的資訊，但本書卻不同了，本書活用了書籍的不變特性，同時保留了這個特性，這也就是說本書所收集的，都是一些不被網路使用的標題，這樣看起來本書好像沒什麼價值。

內容

讓我興起收集網路標題並製作成書，是在著手製作本書的２年前，即1997年春，我參考了各種網路雜誌並企劃「標題競賽」，來收集海內外各網站、企業產品CM等等的標題，另外我還搜尋零星出現在引擎網路雜誌上的標題。而本書的主旨，除了讓讀者對日本美少女CG作家網站有更深一層的了解之外，還希望各位讀者能將此書內容運用在其他網站（書籤集）上，且原則上我們並不主張改變企業的CM標題，不過就CG網站而言，看到的只是網路標題圖樣，因此作者必須清楚地標示出圖樣是出自什麼樣的網站，以及各位讀者能看到什麼樣的CG，而基本上，所謂的標題，是針對規格較小的CG，運用自己所有的技巧來表現個人風格以吸引讀者的注意。

本書內容為各種網路標題，但究竟是什麼樣的標題呢？「簡而言之，就是有關網站中網頁情報畫像」以及「CG短句」。

另外有關時下雜誌的企劃，在CG網站上也是相當流行，而作者不僅竭盡腦汁介紹其特殊風格，同時我們也收集了許多企業的CM標題，至於CG網站中標題作者所流傳下來的獨特思維作品，如今已經很少見了。

標題作品

本書是以收集網路標題作品為主的書，我想今後不只是大受歡迎的CG網站會有網路標題的「作品集」，它已儼然成為日本獨特的藝術文化。

在偶然的機會下，我有幸能和畫刊社的編輯一起討論有關書籍規劃的相關事宜，我將前些日子準備好的企劃方向，大略地向他們說明了一下，而在會談結束後的閒談中，我們決定了本書的命名「標題作品集」，進而開始了更具體的會談。

結果本書便於焉誕生，就圖形而言這是一本相當艱深的書，關於這一點在後序中，將有詳盡的說明討論。

為什麼是這種規格？

本書規格為200位元x40位元，這是目前日本CG作家網站最普遍使用的規格，這種規格足以容納我們收集的網路標題，此標題規格從以前就很受人注目，至於專職整合網路標題的「標題俱樂部（http:www.din.or.jp/rin/bnnr/）」經營者Woody-RINN小姐也表示「希望其在自然而然中，成為日本國內繪畫網站的標桿」，由於起源已很難探尋，我想此種規格之所以能如此被廣泛使用，Woody-RINN小姐居功厥偉。

Basic Concept

The banners in this book are in my mind works of art. They are fundamentally Web sites where artists exhibit their original CG work. I envisaged a collection of banners that would entertain readers and make them respond appreciatively with "That's beautiful!" or "That one's cute!" or "Wow!" A book that people would be glad to have. Thumbing through the pages of a book in your hand iș a different kind of enjoyment to simply looking at banners on a Web link page. Books about the Internet are usually out-of-date the moment they are put on the market. Whereas, I wanted to make the most of the never-changing written word and produce a book that would still be significant far into the future. Indeed I think that this book will increase in value as the banners published here stop being used on the Internet.

How the book began

I first had the idea of compiling a collection of banners in a book in the spring of 1997, about year before I actually started on it. I was in charge of setting up/planning several pages featuring a 'banner contest' for an Internet magazine. At that time the idea of banners from foreign Web sites was catching on and they started popping up here and there advertising companies and their products on search engines and Web magazines in Japan.

Around the same time, banners were already beginning to permeate into the Web sites for CG artists of the bishoujo genre (animation/manga featuring beautiful young teenage girls). In much the same way as commercial banners, they were used in link pages of other Web sites (bookmarks) to entice people to click onto their sites. But with most CG sites, a look at the picture contained on the banner shows you what kind of site there is in store if you visit the homepage and what computer graphics can be seen there. In this way, basically banners of this genre convey the creative power of the individual artists who use the full range of their techniques condensed in the small-sized CG.

In this book I asked the creators of the banners featured here to describe "banner"? It was defined as "screen data concisely showing information relating to the home page it is linked to" and as "CG haiku."

When I was working on the pages for the magazine's banner contest, I remember how extremely difficult it was trying to get a decent collection of advertising banners together despite the huge boom in CG sites. It was not as if I could specialize in bishojo for this job. I could not find any commercial banners that were as creative/expressive as those of the CG Web sites.

Works of Art

It was soon after this that I thought it would be good to have a book featuring artistic banners, a collection of artworks comprising solely of CG Web site banners that are all the rage now. And to present this genre of banner as an original form of Japanese pop art/culture.

I happened to see the editor of Graphic-sha Publishing Co., Ltd. to discuss the planning of a certain book. After a meeting during which I explained the time-consuming preparation the planning involved, we ended up chatting about my ideas for the Banner Collection. That set the ball rolling.

And as you see, the chat resulted in this book. It turned out that the book was extremely difficult to compile. I have written on the problems I experienced in the process in the postscript.

Why this size?

The banners featured in this book are 200 x 40 pixels. I use this size because predominantly that is the size they are on CG artists' Web sites in Japan. And if they were not this size, then I would not have been able to have the large number of banners that I have in my collection.

Early on in the history of the banner, Woody-RINN, who operates Banner Club (http://www.din.or.jp/~rinn/bnnr/), a link site for banner collections, took note of their size. "It seems that artists' sites have naturally taken this as the national standard," It is impossible to know exactly how this happened, but Woody-RINN should take a lot of the credit for its popularization.

CONTENTS

前言	2	**Introduction**	2
例言	7	**Editorial Notes**	7
標題及CG作品	8	**Banners and CG Artworks**	8
索引	124	**Index**	124
後序	142	**Afterword**	142

CATEGORIES

ORIGINAL 主要是獨創性作品以及網址	8	**ORIGINAL** Sites featuring mainly original characters	8
PUNI 登載PUNI可愛女孩的作品網址	62	**PUNI** Cute girl characters with rounded baby faces	62
ANIME/COMIC/GAME 主要是以動畫、喜劇及電玩特性為題材的作品，以及相關作品網址	64	**ANIME/COMIC/GAME** Sites featuring or about anime, comic or game characters	64
FANTASY 主要是以幻想世界為題材的作品以及作品網址	92	**FANTASY** About or featuring mainly fantasy worlds	92
3D 登載3D畫集的網址	98	**3D** Sites focusing on 3D graphics	98
COSTUME 登載相關服飾作品的網址	100	**COSTUME** Images of characters wearing specially designed costumes	100
KEMONO 登載著屬於幻想世界中的身體一部份有貓耳以及尾巴特性為題材的作品網址	106	**KEMONO** Beast girls such as cat girls or other part – animal fantasy characters	106
SOFTWARE 登載發表相關CG效應的網址，發表擁有CG重要意境的電玩網址等等	110	**SOFTWARE** Sites presenting CG utility programs or computer graphic games	110
ADULT 登載成人及藝術性的網址	112	**ADULT** Adult-oriented artistic images	112
LINK/INFORMATION/OTHERS 作品集、情報站、小說、其他	118	**LINK/INFORMATION/OTHERS** Link collections, information sites, online novels, etc.	118
SPECIAL 特殊網址及已截止的網站	122	**SPECIAL** Special last minute addition sites	122

種類

因作者採用的是主觀區分法，若有與網站作者意向相異之處，敬請見諒。標題則沿用先前網羅的資訊，而為了編輯進度，並未對相關種類進行詳細的比對區分，而在編集後，又因方針改變，附錄部份的區分可能不夠明確，故請各位以直覺來分析網路標題。

範例

1）標題作品資料的刊載順序如下：

標題圖片
網址
網站名/作者名

2）對於登載的網路標題及CG，超熱門部份以(C)標示出，但即使沒有標示以(C)的作品，作品的權利者或作者，也保有同等的權利。

3）對本書登載的標題及CG作品，因大部分都沒有考慮到紙質印刷的特性，若在閱讀中顏色有任何異常現象，還請見諒。

Categories

The web sites have been divided into the following categories at my own discretion so I apologize if these categories go against the web authors' ideas and for their vaguness. These categories are only suggestions however so the viewers should make up their own minds.

Editorial Notes

1)Data on each banner artwork is in the following order;

The banner picture
The URL of the site
The name of the site/creator

2)About the Copyright Mark

I have entered a copyright mark for the banners and CG featured in the book only where specifically requested. The rights of those whose works do not have a mark, their copyrights are protected too.

3)Printed Reproduction of the Work

Most of the banners and CG artwork featured in the book were never intended for paper media. Therefore the color quality may be inferior to that appreciated on the computer screen.

http://www.age.ne.jp/x/fam/
爆熱FAM_Activate( ) / FAM柳瀬

http://www.age.ne.jp/x/fam/
爆熱FAM_Activate( ) / FAM柳瀬

http://www.ainet.or.jp/~cue/
全国庁 にこりん情報局 / cue

http://plaza26.mbn.or.jp/~chorop/
ぼこぼこダイナマイツ / Chororphis

http://plaza26.mbn.or.jp/~chorop/
ぼこぼこダイナマイツ / Chororphis

http://www02.so-net.ne.jp/~hiropon/
元気薬 / HIRO PON

http://home.intercity.or.jp/users/K/
KENホームページ / KEN

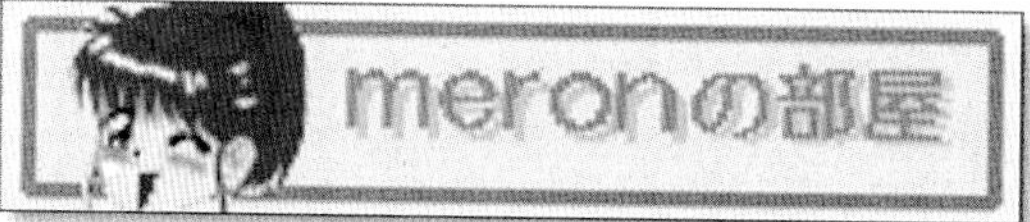

http://plaza4.mbn.or.jp/~meron/
meronの部屋 / meron

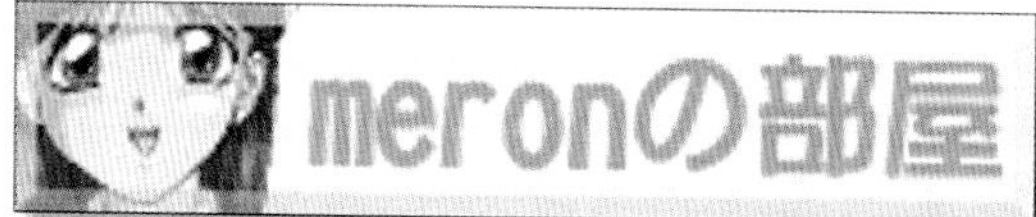

http://plaza4.mbn.or.jp/~meron/
meronの部屋 / meron

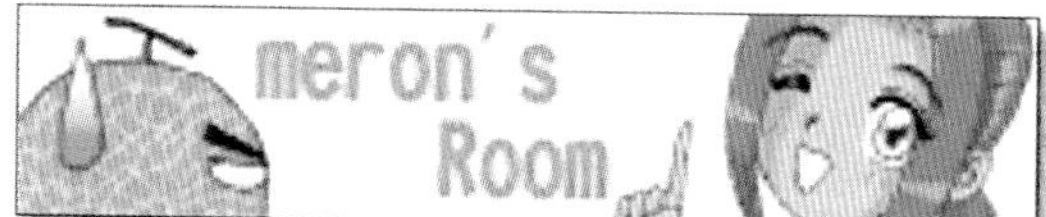

http://plaza4.mbn.or.jp/~meron
meronの部屋 / meron

http://plaza4.mbn.or.jp/~meron
meronの部屋 / meron

http://plaza4.mbn.or.jp/~meron
meronの部屋 / meron

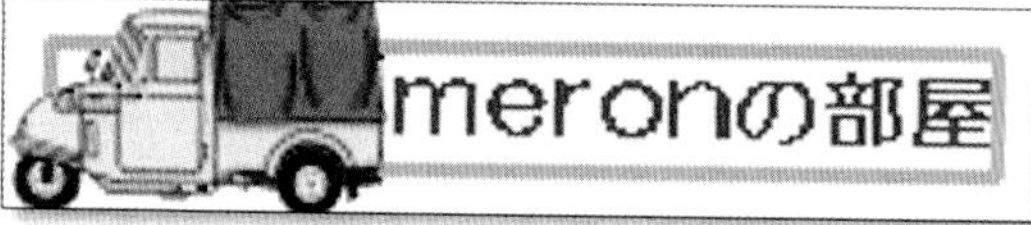

http://plaza4.mbn.or.jp/~meron
meronの部屋 / meron

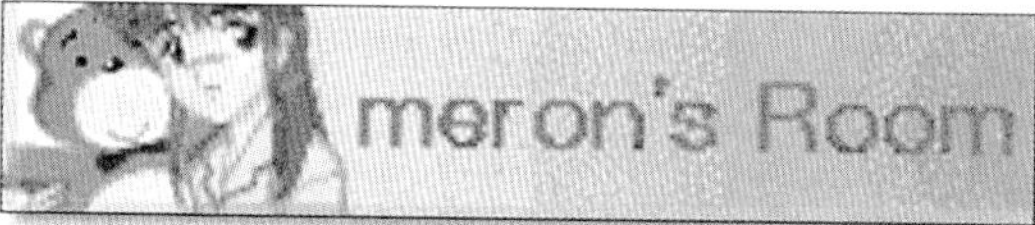

http://plaza4.mbn.or.jp/~meron
meronの部屋 / meron

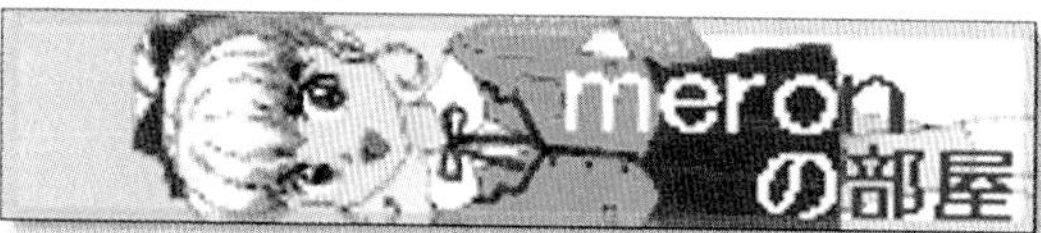

http://plaza4.mbn.or.jp/~meron
meronの部屋 / meron

http://plaza4.mbn.or.jp/~meron
meronの部屋 / meron

http://plaza4.mbn.or.jp/~meron
meronの部屋 / meron

http://plaza4.mbn.or.jp/~meron
meronの部屋 / meron

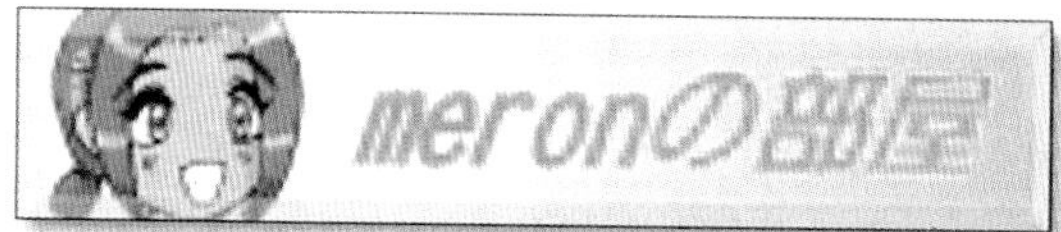

http://plaza4.mbn.or.jp/~meron
meronの部屋 / meron

http://www7.big.or.jp/~kft/NEKO/index.html
猫族集会 / KFT

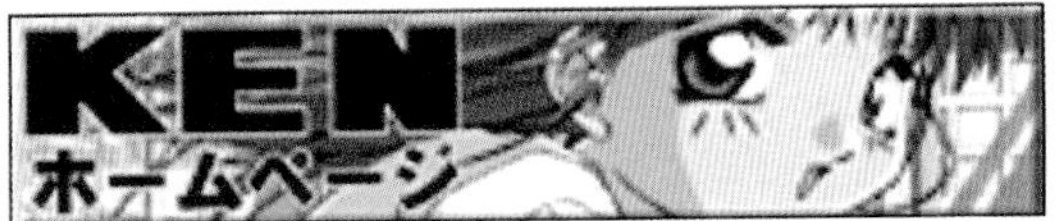

KENホームページ
http://home.intercity.or.jp/users/K/
KEN

可欣賞到許多精神奕奕女孩的CG，其有遊戲卡和電玩作品，而其共通點是，不論是哪一個PANCHIRA作品，都是健康又可愛的，所以不要害怕被人稱之為PANCHIRA作家，請以世界第一的PANCHIRA作家為奮鬥目標，我會替你加油的。

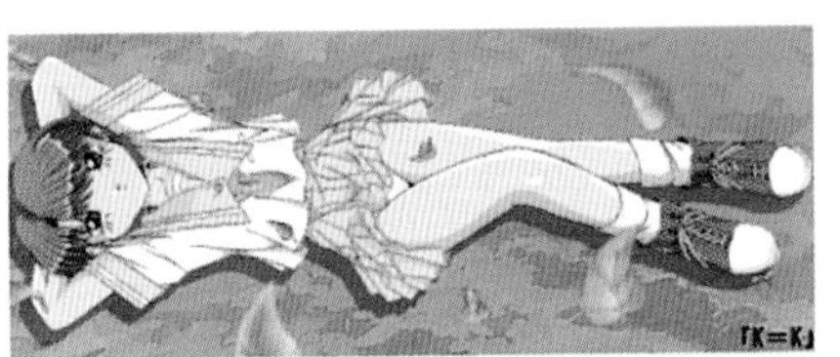

This site features many images of genki lively girls including popular anime and game characters. The special theme of these images is panchira. This word is a combination of the words "panty" and "chira chira" which is the sound effect for a brief glance or flash. As you can imagine, there are many images of mini-skirted girl characters. The images are very cute and in my opinion, present an innocent and healthy view of sexuality. Although the author of this site prefers not to be refered to as a master of panchira, I think he should accept the title with pride and strive to become the panchira master of the world.

http://www.cc.rim.or.jp/~mokkun
Alisa The Wonderland / Mokkun

http://www.cc.rim.or.jp/~mokkun
Alisa The Wonderland / Mokkun

http://plaza11.mbn.or.jp/~sekiya/dsean/
D-SeaNET / sachi

http://village.infoweb.ne.jp/~fwix7358
非万能電化研究所 / miho

http://www.ceres.dti.ne.jp/~miwaza
GUILD HOUSE/ 神之みわざ

http://www.vector.co.jp/authors/VA008796/
くろうさ工房 / oku

http://www.alles.or.jp/~nayu
QUEERNESS / Nayu

http://plaza29.mbn.or.jp/~mururu/index.htm
R's ROOM「落描きのススメ」/ MRR

http://plaza29.mbn.or.jp/~mururu/index.htm
R's ROOM「落描きのススメ」/ MRR

http://plaza29.mbn.or.jp/~mururu/index.htm
R's ROOM「落描きのススメ」/ MRR

http://home4.highway.ne.jp/ramble/index.htm
ramble's home page / ramble

http://www.asahi-net.or.jp/~mt3t-njr
Rosanna's Homepage / Rosanna

http://www.asahi-net.or.jp/~mt3t-njr
Rosanna's Homepage / Rosanna

http://www.asahi-net.or.jp/~mt3t-njr
Rosanna's Homepage / Rosanna

http://cf.vow.co.jp/rss
ルートS.S.のハッピーカムカム / Route S.S.

http://www.asahi-net.or.jp/~FP7Y-OONS
するめや / Oh-Shall & TEN

http://www.asahi-net.or.jp/~FP7Y-OONS
するめや / Oh-Shall & TEN

http://plaza3.mbn.or.jp/~pota
Web Potahouse / SAS_P

http://plaza3.mbn.or.jp/~pota
Web Potahouse / SAS_P

http://www.ktroad.ne.jp/~ryota
<蔀>しとみ / RYOTA

GUILD HOUSE

GUILD HOUSE
http://www.ceres.dti.ne.jp/~miwaza
神之みわざ

有如繪畫般的細緻作品，此類作品都是從故事中截取一部分來完成的，畫中房子及人物背景都相當富有想像力喔！
一定要看喔！作者CG的毛髮描繪手法解說網頁。

Beautifully detailed and realistically portrayed scenes are this site's specialty. Each image is like a scene taken from a story and encourages you to imagine the background behind the characters depicted and the environments they live in. There is also a section featuring a detailed step by step of how to draw anime-style girl's hair. This is a must-see.

ORIGINAL

http://www.asahi-net.or.jp/~QM4H-IIM/index.htm
Yeemar's Home Page / Yeemar

http://www.asahi-net.or.jp/~QM4H-IIM/index.htm
Yeemar's Home Page / Yeemar

http://www.bekkoame.ne.jp/~y.shiro
SHIROのお絵書き工房 / SHIRO

http://www.din.or.jp/~alger
Algernon's Artless Arts / Algernon

http://www.din.or.jp/~alger
Algernon's Artless Arts / Algernon

http://www.din.or.jp/~alger
Algernon's Artless Arts / Algernon

http://www.din.or.jp/~rinn
RINN's Page / Woody-RINN

http://www.din.or.jp/~rinn
RINN's Page / Woody-RINN

http://www.din.or.jp/~rinn
RINN's Page / Woody-RINN

http://www.din.or.jp/~rinn
RINN's Page / Woody-RINN

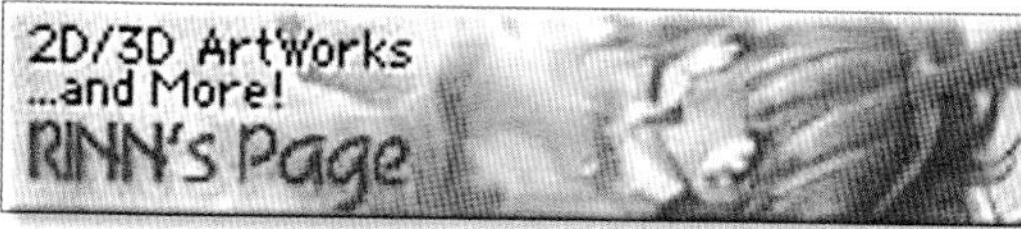

http://www.din.or.jp/~rinn
RINN's Page / Woody-RINN

http://www.din.or.jp/~rinn
RINN's Page / Woody-RINN

http://www.din.or.jp/~rinn
RINN's Page / Woody-RINN

http://www.alphatec.or.jp/~shouta
SHOU's GALLERY / sho

http://member.nifty.ne.jp/seto
Seto Home Studio / Seto

http://www.ask.or.jp/~takam
Takam 作品集 / Takam

http://www.alles.or.jp/~takefour
FREEZE MOON / Take-Four

http://w3.mtci.ne.jp/~tash
SYNTHESIS / TASH

http://cgi.din.or.jp/~ta2chi4/dream
夢戦士ドリームソルジャーZ / System 3(ag)

http://www.intacc.ne.jp/~gisho
ZX-RR CG PAGE / Rai

2D/3D ArtWorks
...and More!
RINN's Page

RINN's Page

http://www.din.or.jp/~rinn

Woody-RINN

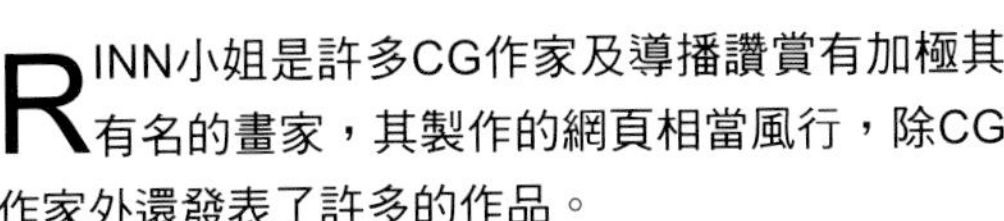

RINN小姐是許多CG作家及導播讚賞有加極其有名的畫家，其製作的網頁相當風行，除CG作家外還發表了許多的作品。

Rinn is a very well-known artist who is highly respected by other CG artists and also programmers. His home page gives tips and advice on all the necessary elements for creating good web pages including HTML tips and new ideas for web designs. He also runs the "Banner Club". This started as a small corner of his main home page to display his collection of anime web banners and was then expanded into a complete site in its own right. Banner Club was responsible for helping to popularize and set the current 200x400 pixel standard for anime web banners such as the banners featured in this book.

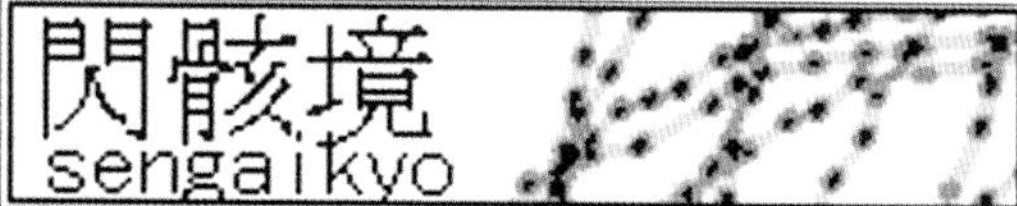

http://village.infoweb.ne.jp/~fwie0190
閃骸境 / ありさわとよみつ

http://w3ma.kcom.ne.jp/~k-ogawa/charm
ちゃ〜むなねっと / You

http://w3ma.kcom.ne.jp/~k-ogawa/charm
ちゃ〜むなねっと / You

http://w3ma.kcom.ne.jp/~k-ogawa/charm
ちゃ〜むなねっと / You

http://w3ma.kcom.ne.jp/~k-ogawa/charm
ちゃ〜むなねっと / You

http://w3ma.kcom.ne.jp/~k-ogawa/charm
ちゃ〜むなねっと / You

http://www.bekkoame.ne.jp/~woodstok/yui
GRAVITATION / かずきゆい

http://www.bekkoame.ne.jp/~woodstok/yui
GRAVITATION / かずきゆい

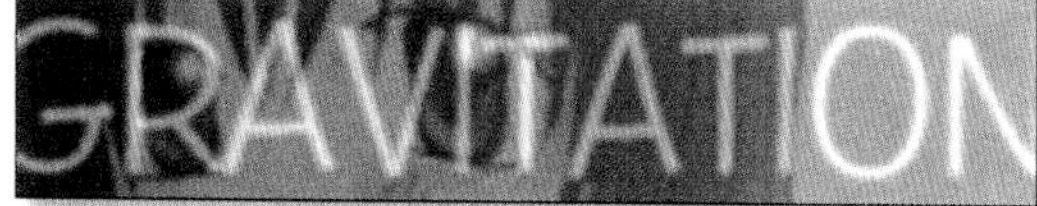

http://www.bekkoame.ne.jp/~woodstok/yui
GRAVITATION / かずきゆい

http://www.bekkoame.ne.jp/~woodstok/yui
GRAVITATION / かずきゆい

http://www.bekkoame.ne.jp/~woodstok/yui
GRAVITATION / かずきゆい

http://www.bekkoame.ne.jp/~woodstok/yui
GRAVITATION / かずきゆい

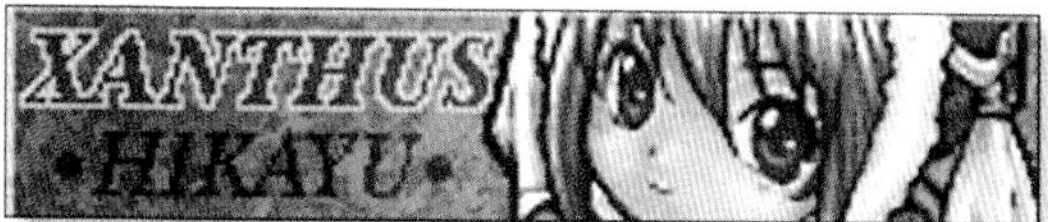

http://www.din.or.jp/~hikayu
XANTHUS / XANTHUS

http://www.din.or.jp/~hikayu
XANTHUS / XANTHUS

http://www.asahi-net.or.jp/~nx1m-stu
工房猫の手 / あらきよう

http://home.highway.or.jp/yosshi/main/kagome.htm
PHANTASYSTEM / かごめ

http://www.age.ne.jp/x/tadashi
without sun /あいね

http://www.top.or.jp/~xanthus
XANTHUS & PAPILLON / XANTHUS & PAPILLON

http://aumy.biwako.shiga-u.ac.jp/std/s96762hw/youmei.htm
YOUMEI's home page / Youmei

XANTHUS
http://www.din.or.jp/~hikayu
XANTHUS

喜歡有點剛毅眼神以及感覺纖弱的下巴線條，HIKAYU小姐為職業插畫家，她同時還活躍於雜誌的領域。

Characacters like these with bright intense eyes and faces with youthful rounded chins are my favorite. Hikayu is a professional illustrator who regularily contributes illustrations to magazines.

今年也請多多賜教。

http://www.cc.rim.or.jp/~momoko
キラキラ★ヒカル! / さっぽろモモコ

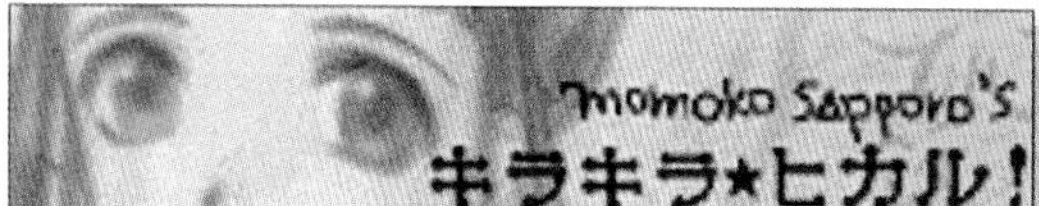

http://www.cc.rim.or.jp/~momoko
キラキラ★ヒカル! / さっぽろモモコ

http://plaza26.mbn.or.jp/~chikamichi
近道。/ かせん

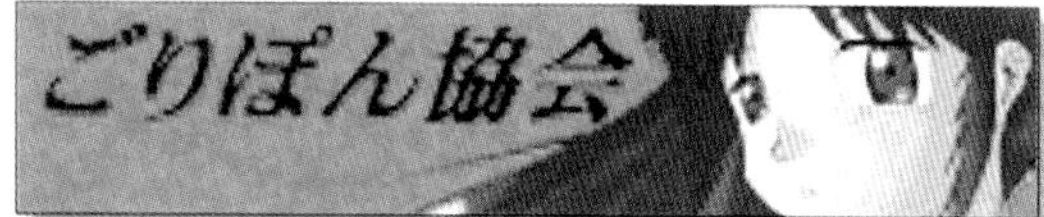

http://www.ask.or.jp/~seishiro/goripon.htm
ごりぽん協会ホームページ / ごりぽん協会

http://super.win.ne.jp/~gato
がとーのStudioBEYOND / がとー

http://fame.calen.ne.jp/~kawachi
目薬専門店ねぼけ堂 / かわち丸

http://fame.calen.ne.jp/~kawachi
目薬専門店ねぼけ堂 / かわち丸

http://www.bekkoame.ne.jp/~kch
Mystic Kingdom / こちこち

http://www.bekkoame.ne.jp/~kch
Mystic Kingdom / こちこち

http://home3.highway.ne.jp/kattin
Monotone Club' / かっちん

http://home3.highway.ne.jp/kattin
Monotone Club' / かっちん

http://www.crt.or.jp/public/user/~yama
YAMA's HomePage/ YAMA

http://www.asahi-net.or.jp/~iu7k-eczn
しゅう・とくとみの忍者貴族 / しゅう・とくとみ

http://www.asahi-net.or.jp/~iu7k-eczn
しゅう・とくとみの忍者貴族 / しゅう・とくとみ

http://hiroba.net/scratch
SCRATCH / たかぼ

http://www.big.or.jp/~t-shun/index.html
はるのつぼみ / つぼみしゅん

http://www.amy.hi-ho.ne.jp/takamichi
TAKAMICHI FACTORY / たかみち

http://plaza12.mbn.or.jp/~takasiro
たかしろ亭 / たかしろそうま

http://www.atnet.ne.jp/~toshit
Toshi-B's image Factry / トシB

http://www.atnet.ne.jp/~toshit
Toshi-B's image Factry / トシB

YAMA's HomePage
http://www.crt.or.jp/public/user/~yama
YAMA

YAMA

YAMA

YAMA小姐的網站登載的是具「真實感」的作品，在第一次拜訪這個網站時，我看到了真實描畫的女子，我終於有機會與理想女性相會，網站中還有「卡通動畫性」作品，而且不論是哪一個作品都很可愛，所以請實際上網來看一看。

These are some examples of the highly realistic CG portraits you can see if you visit Yama's site. When I first visited this site and saw one of Yama's very realistic portraits, I felt as if I had met my ideal woman. In addition to these realistic images, cute anime-style images are also displayed so please visit and have fun admiring the beautiful images of this site.

http://www.din.or.jp/~beershop
麦酒店舗 / ぱぴよん

http://home.intercity.or.jp/users/onishi
めいへう゛ん / ぼいどめいん

http://plaza8.mbn.or.jp/~nonn
らいとっ! / のん

http://plaza8.mbn.or.jp/~nonn
らいとっ! / のん

http://www.ask.or.jp/~byte
東風 / ばいと屋某

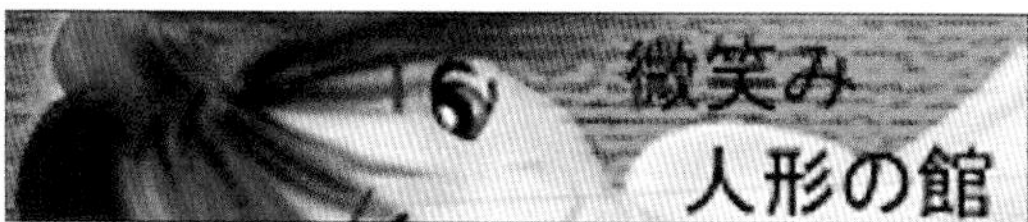

http://www4.justnet.ne.jp/~dolls
微笑み人形の館 / ひの

http://www4.justnet.ne.jp/~dolls
微笑み人形の館 / ひの

http://home.interlink.or.jp/~hinata
喫茶「ひだまり」 / ひなた☆すう

http://club.pep.ne.jp/~hyoi.kaz/
Hyoi's House Leaves / ひょい

http://www-user.interq.or.jp/~nan
水精少女譚 / ななみ

http://www.annie.ne.jp/~miura
BLACK STARS / みうらたけひろ

http://nanami.px.to
天体少女図鑑 / ななみ

http://nanami.px.to
天体少女図鑑 / ななみ

http://nanami.px.to
天体少女図鑑 / ななみ

http://csefs01.ce.nihon-u.ac.jp/~u086119
のにさんのホームページだ! / のにさん

http://csefs01.ce.nihon-u.ac.jp/~u086119
のにさんのホームページだ! / のにさん

http://www.ap.kyushu-u.ac.jp/appphy/mer
まさあきのCGギャラリー / まさあき

http://netv5.netvision.co.jp/~pi
Pi's HomePage / ぴー

http://netv5.netvision.co.jp/~pi
Pi's HomePage / ぴー

http://netv5.netvision.co.jp/~pi
Pi's HomePage / ぴー

Hyoi's House Leaves

http://club.infopepper.or.jp/~hyoi.kaz

ひょい

網站佣人、「身帶頸圈」非常可愛，具有雙重角色特性，這是「漫畫企畫屋」的一角，有趣的是這網站有自己的創意空間，並利用通信傳達給讀者，這在漫畫中是很少見的，但卻很有趣。

A pretty character named "Puni Robo-maid" leads you to each page of the site. One of the most interesting sections of this site is the Manga Planning Room, an interactive page where the artist and visitors collaborate together to produce original manga, exchanging their ideas and suggestions. This section was planned specifically to take advantage of the interactivity of the internet.

ORIGINAL

http://www2s.biglobe.ne.jp/~JEWEL
JEWEL MASTER / 結城 梓

http://home3.highway.ne.jp/nisa
えりくしる / 紅竜鷹羽

http://home.interlink.or.jp/~chiharu
enocchy's ROOM / 榎本 千春

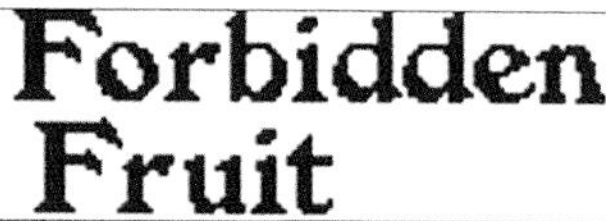

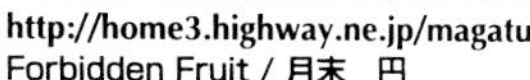

http://home3.highway.ne.jp/magatu
Forbidden Fruit / 月末　円

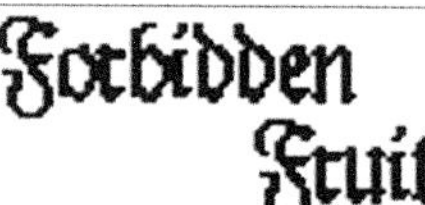

http://home3.highway.ne.jp/magatu
Forbidden Fruit / 月末　円

http://member.nifty.ne.jp/hitode
海星★梨の落書きホームページ / 海星★梨

http://w32.mtci.ne.jp/~muho
Muhoho's Homepage! / むほほ

http://www.bekkoame.ne.jp/~iiyoshi
天然色 / 井上海松

http://www.246.ne.jp/~reiko-h
猫秘密情報結社 / 岩崎れえこ

http://www.246.ne.jp/~reiko-h
猫秘密情報結社 / 岩崎れえこ

http://www.246.ne.jp/~reiko-h
猫秘密情報結社 / 岩崎れえこ

http://village.infoweb.ne.jp/~fwhe8793
ぺぱーMintグリーンホームページ / 安木美恵子

http://www.alles.or.jp/~kurage
Soft Page ～やわらかいものすきですか?～ / ミズクラゲ

http://member.nifty.ne.jp/yoshinoya
特盛EXPRESS / 吉野家うっしー

http://www.asahi-net.or.jp/~ty5a-kmr
Rakia CG Garden / ラキア

http://www.annie.ne.jp/~meka
化け物アイランド / メカうほほ1号

http://plaza6.mbn.or.jp/~YGF
Y's Graphic factory / ワイズ

http://www.osk.3web.ne.jp/~yukino/（現在更新停止中）
娘細工 / ゆきの

http://plaza19.mbn.or.jp/~zen_yasumori
Zero Second / 安森 然

海星★梨の落書きホームページ
http://member.nifty.ne.jp/hitode/
海星★梨

HITODE★NASHI 1998

以漢字和符號所組成的筆名，讀作「海星★梨」，卡通CG是以畫線方式來著色的，而完成作品一般也都會殘留一些畫線，可是海星★梨的作品卻不曾出現這種現象，而且色彩及對比分明，非常獨特。

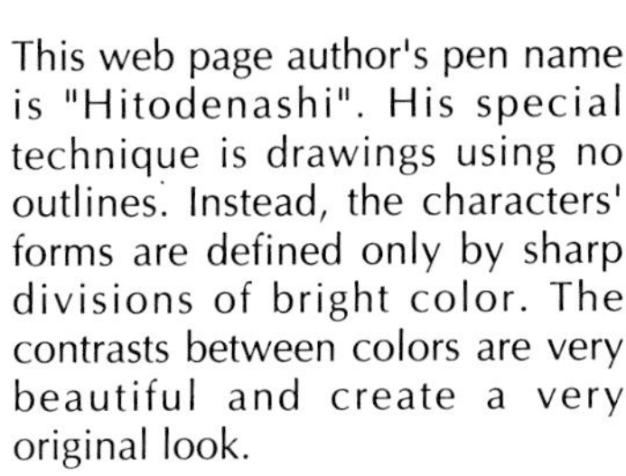

This web page author's pen name is "Hitodenashi". His special technique is drawings using no outlines. Instead, the characters' forms are defined only by sharp divisions of bright color. The contrasts between colors are very beautiful and create a very original look.

ORIGINAL

http://www.din.or.jp/~akiduki
ありすの喫茶店 / 秋月あかね

http://www.bekkoame.ne.jp/ha/sagiri
プリンシュー / 須田さぎり

http://www.246.ne.jp/~c-sakura
佐倉千歳堂 / 佐倉千歳

http://www.asahi-net.or.jp/~JL6T-NKGW
ひまわりカンパニー / 七瀬いーうぃ

http://www.asahi-net.or.jp/~JL6T-NKGW
ひまわりカンパニー / 七瀬いーうぃ

http://www.asahi-net.or.jp/~JL6T-NKGW
ひまわりカンパニー / 七瀬いーうぃ

http://ha3.seikyou.ne.jp/home/kouyou
ORIENT KOUYOU / 荒葉

http://ha3.seikyou.ne.jp/home/kouyou
ORIENT KOUYOU / 荒葉

http://www.aland.to/~e-ksato
佐藤和芳アートギャラリー / 佐藤和芳

http://www.bekkoame.ne.jp/~tettete
Ah Mitsuketa! / 高橋哲人

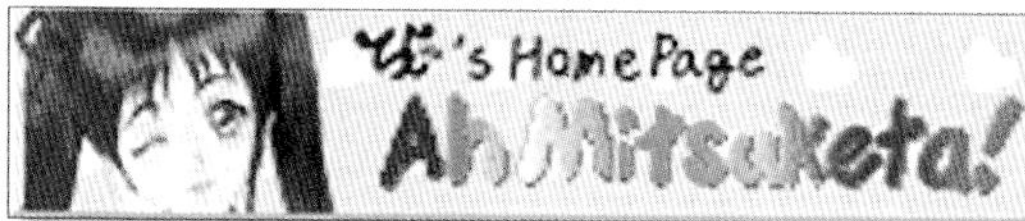

http://www.bekkoame.ne.jp/~tettete
Ah Mitsuketa! / 高橋哲人

http://www.246.ne.jp/~miz
しゃんふぁ・なにむ / 水明&瑞浪かさね

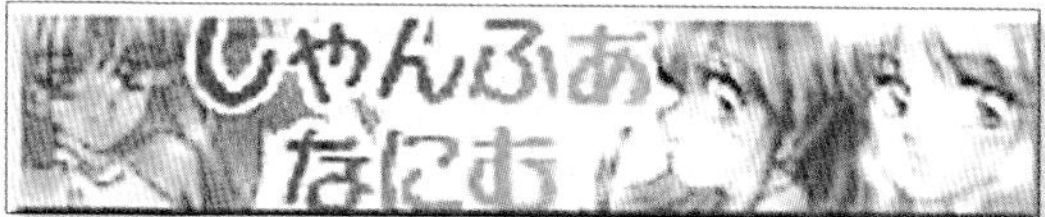

http://www.246.ne.jp/~miz
しゃんふぁ・なにむ / 水明&瑞浪かさね

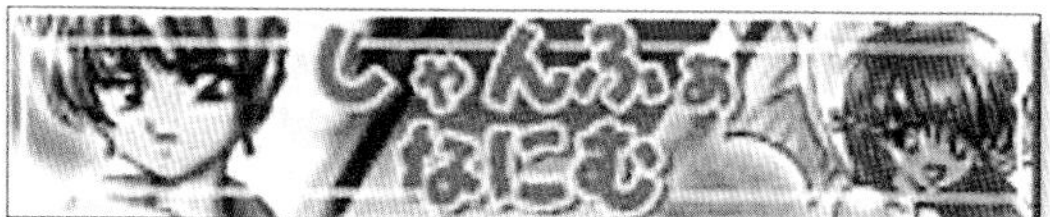

http://www.246.ne.jp/~miz
しゃんふぁ・なにむ / 水明&瑞浪かさね

http://www.246.ne.jp/~miz
しゃんふぁ・なにむ / 水明&瑞浪かさね

http://www.246.ne.jp/~miz
しゃんふぁ・なにむ / 水明&瑞浪かさね

http://village.infoweb.ne.jp/~fwgi0240
小田修紀CG館 / 小田修紀

http://www.246.ne.jp/~ritzve
Ritzve's Dreamscape / りつべ

http://www.246.ne.jp/~ritzve
Ritzve's Dreamscape / りつべ

http://www.bekkoame.ne.jp/~keen/yuna
香月柚奈 ほーむぺーじ・こざるのゆなちゃん / 香月柚奈

因作畫者及配樂者都是相當頂級的廣告製作者，RITSUBE小姐將所謂的「DeBabelizer」運用在縮小的CG上，其是利用Photoshop的雙重立方縮小算式來縮小的，所以就算失敗了，還是看起來美美的，我也曾嘗試過這個方法，確實感覺不同，詳細情形請上網查看。

http://www,debabelizer.com.

A professional graphic and sound creator, Ritzve uses the program DeBabelizer to compress all the images for his web pages. Using DeBabelizer's "Sign Sharp" setting when reducing the size of images is superior to Photoshop's "Bi-cubic" setting which can sometimes distort the image. I've tried DeBabelizer myself also and I can see a noticable difference from Photoshop. For more information on DeBabelizer, please visit: http://www.debabelizer.com/

ORIGINAL

http://www.din.or.jp/~abara
ARTCH / 宮村和生

http://www.din.or.jp/~abara
ARTCH / 宮村和生

http://web.kyoto-inet.or.jp/people/aoshima
新あおしま製菓 / 青島ういろう

http://www2u.biglobe.ne.jp/~hinato
ひろいずむ / 大鳴ひなと

http://plaza25.mbn.or.jp/~junchiro
八尋殿 / 朝倉純壱

http://home.intercity.or.jp/users/accell
嗚呼! 極限劇場 / 高木裕司

http://home.intercity.or.jp/users/accell
嗚呼! 極限劇場 / 高木裕司

http://home.intercity.or.jp/users/accell
嗚呼! 極限劇場 / 高木裕司

http://home.intercity.or.jp/users/accell
嗚呼! 極限劇場 / 高木裕司

http://home.intercity.or.jp/users/accell
嗚呼! 極限劇場 / 高木裕司

http://home4.highway.ne.jp/mizuho/how
Bliss and Rapture... / 瑞穂

http://nug.nasu-net.or.jp/~tmakino
Moonlight Aquarium / 川原 美紀

http://plaza15.mbn.or.jp/~sansyou
POINT BLANK / 山椒 奈々実

http://plaza15.mbn.or.jp/~sansyou
POINT BLANK / 山椒 奈々実

http://www.din.or.jp/~syuriyan
MEI-Q-RONDO / 珠梨やすゆき

http://www.big.or.jp/~tomoki
TOMORAの生けどりInternet / 智羅

http://www.geocities.co.jp/Playtown-Denei/8325/
ZANY / 倉本なつね

http://www.geocities.co.jp/Playtown-Denei/8325/
ZANY / 倉本なつね

http://www.geocities.co.jp/Playtown-Denei/8325/
ZANY / 倉本なつね

http://www.geocities.co.jp/Playtown-Denei/8325/
ZANY / 倉本なつね

嗚呼！極限劇場
http://home.intercity.or.jp/users/accell/
高木裕司

因為喜歡所謂的「大眼妹妹」而登載於此，而我也很喜歡「胖嘟嘟的臉蛋」。

As I really like girls with the kind of eyes featured here (widely spaced, down-turning), I especially requested the artist to allow me to publish his images in my book. I also like "puni" which is also the style of the girl characters here.

ORIGINAL

http://www.asahi-net.or.jp/~qn9m-ysd/ammonite/
あんも・ないとの部屋 / あんも・ないと

http://member.nifty.ne.jp/devilkitten/toppage.html
暗黒太陽通信 / 藤川純一

http://w3.mtci.ne.jp/~ayamaru/
綾丸堂本店 / 龍王綾丸

http://www.asahi-net.or.jp/~cz6t-sm/
迷宮美術館 / 夕凪雄那

http://www.asahi-net.or.jp/~cz6t-sm/
迷宮美術館 / 夕凪雄那

http://www.asahi-net.or.jp/~jp8k-iwsk
ねこわんこホームページ / ねこねこ

http://www.bekkoame.ne.jp/i/fujisaki/
うさうさFのホームページ / うさうさF

http://www.bekkoame.ne.jp/i/fujisaki/
うさうさFのホームページ / うさうさF

http://home3.highway.ne.jp/h-mikoto
はにわ南蛮 / 花形水琴

http://home3.highway.ne.jp/h-mikoto
はにわ南蛮 / 花形水琴

http://w3.mtci.ne.jp/~hotaru-k/
ほたる狩り / 恋純☆ほたる

http://www.blk.mmtr.or.jp/~shuuhei/
木曽屋本店 / 木曽秀平

http://www.dd.iij4u.or.jp/~lesia/index.html
まじかるすてーしょん / LESIA

http://www.bekkoame.ne.jp/ha/masako/
秘書室インターネット版 / 秘書

http://ha1.seikyou.ne.jp/home/yuunagi
夕凪堂本舗 / 東雲あずみ

http://www.asahi-net.or.jp/~jp8k-iwsk
ねこわんこホームページ / ねこねこ

http://www.asahi-net.or.jp/~jp8k-iwsk
ねこわんこホームページ / ねこねこ

http://www.asahi-net.or.jp/~jp8k-iwsk
ねこわんこホームページ / ねこねこ

http://www.asahi-net.or.jp/~jp8k-iwsk
ねこわんこホームページ / ねこねこ

http://www.asahi-net.or.jp/~jp8k-iwsk
ねこわんこホームページ / ねこねこ

はにわ南蛮

http://home3.highway.ne.jp/h-mikoto/

花形水琴

作品統一為淡色的網站，設計風格統一，使用的界面卡也是統一規格，其可分成卡通性和設計性 2 種畫風，而作者我也被「CAPSUL DOLL」所迷惑。

The use of pastel colors throughout and the consistent interface combine in making this a very well-designed site. A combination of anime-style graphics and abstract designs are used together in a complementary style creat-ing a very sophisticated look. I especially liked a particular image here called "Capsul Doll".

http://www.w-w.ne.jp/~1inch/
1inch / 朝妻天

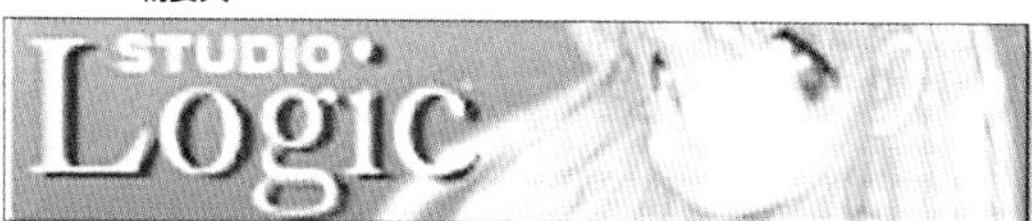

http://www.asahi-net.or.jp/~gu7m-myk/
Studio LOGIC / PANTO MAIMU

http://www.blk.mmtr.or.jp/~itsuki/
Mei Itsuki's Web Page / 伊月めい

http://members.aol.com/rinrin201/hp/Untitled1.html
SOG club / 氷優きゃあ

http://www.cc.rim.or.jp/~urara/
Nocturne / ウララ

http://www.cc.rim.or.jp/~urara/
Nocturne / ウララ

http://www.cc.rim.or.jp/~urara/
Nocturne / ウララ

http://www.cc.rim.or.jp/~urara/
Nocturne / ウララ

http://www.cc.rim.or.jp/~urara/
Nocturne / ウララ

http://www.cc.rim.or.jp/~urara/
Nocturne / ウララ

http://www.ceres.dti.ne.jp/~n621197/index.htm
NOW WON! / NOW

http://www.din.or.jp/~hisumi/
piano:forte a.s.c. / 氷澄水

http://www.din.or.jp/~hisumi/
piano:forte a.s.c. / 氷澄水

http://www.din.or.jp/~hisumi/
piano:forte a.s.c. / 氷澄水

http://www.din.or.jp/~hisumi/
piano:forte a.s.c. / 氷澄水

http://www.din.or.jp/~hisumi/
piano:forte a.s.c. / 氷澄水

http://www.din.or.jp/~hisumi/
piano:forte a.s.c. / 氷澄水

http://www.bekkoame.ne.jp/~imashiro/
PRECIOUS / ふうきりえ

http://www.din.or.jp/~happy-g/
OFF WORLD / フッタこーじ&ハッピーガヴァン

http://member.nifty.ne.jp/suzuki-ken/
SUZUKEN / 鈴木 健

はにわ南蛮
http://home3.highway.ne.jp/h-mikoto/
花形水琴

色調較明顯的作品也很有趣不是嗎？由於雜誌印刷有脫版現象，所以特地為本書重新製作，真是太感激了。

The large swirled color patterns these images use are very interesting, don' t you think? This is an original image created especially for this book, for which I am very grateful.

http://w33.mtci.ne.jp/~mizuna/
Cafe☆ミぽっとはうす / 葉桜みずな

http://w33.mtci.ne.jp/~mizuna/
Ｃａｆｅ☆ミぽっとはうす / 葉桜みずな

http://www.alles.or.jp/~morobosi/
gnosis net zero / Masaki Shibata

http://www.bekkoame.ne.jp/i/gd5787/top.html
GUIDE MAP / ほづみりや

http://www.bekkoame.ne.jp/i/gd5787/top.html
GUIDE MAP / ほづみりや

http://www.click.or.jp/~cha/
少年グラタン/ 茶柱たつや

http://home2.highway.or.jp/syui/
In My Room / 如月庵

http://member.nifty.ne.jp/asomi/
[仮庵] / 島田朝臣

http://hiei.okuma.nuee.nagoya-u.ac.jp/~itaru/
ITAL GOODS Home Page / Ital Goods

http://www.dd.iij4u.or.jp/~kamiza/
kamizanushi room / 神座主志

http://www.dd.iij4u.or.jp/~kamiza/
kamizanushi room / 神座主志

http://www.click.or.jp/~cha/
少年グラタン / 茶柱たつや

http://www.dmp.co.jp/masumic/
仙道ますみのシタゴコロ / 仙道ますみ

http://ha5.seikyou.ne.jp/home/nayuta/nayuta_gado/
那由他画堂 / 那由他

http://www.din.or.jp/~eoaiu/
楽天主義国 / 藍雨絵緒

http://www.dango.ne.jp/~gfa01575/
おやじん家 / おやじ

http://w33.mtci.ne.jp/~mizuna/
Ｃａｆｅ☆ミぽっとはうす / 葉桜みずな

http://www.catnet.ne.jp/yorozuya/
備後萬屋 / さいとうつかさ

http://www.catnet.ne.jp/yorozuya/
備後萬屋 / さいとうつかさ

http://member.nifty.ne.jp/yutaka_n/
Digital 光画堂 / 二條ゆたか

[仮庵]
http://member.nifty.ne.jp/asomi/
島田朝臣

本書的封面也是使用島田的作品，我覺得在描繪方面，島田非常的棒，就卡通而言，其透光性的表現令人吃驚，且在電腦螢幕上觀賞時色彩更形鮮明。

Asomi also created the cover image for the book. Light is depicted very beautifully in Asomi's images. The effect of the shining light is especially vivid when viewed on the computer screen.

ORIGINAL

http://www.sainet.or.jp/~anze/
Anze's HOMEPAGE OasisRoad / 杏世

http://www02.so-net.ne.jp/~ryu-sein/
APPLICANT / セイン流

http://www.webnik.ne.jp/~pukuten/
ASTRO LINER / pukuten

http://www01.u-page.so-net.ne.jp/ja2/ko_yuuse/
Aggregat WORKS / 悠瀬巧一

http://www01.u-page.so-net.ne.jp/ja2/ko_yuuse/
Aggregat WORKS / 悠瀬巧一

http://www.kisnet.or.jp/bitter/
brilliant street / bitter

http://www.kisnet.or.jp/bitter/
brilliant street / bitter

http://www.lofty-tec.co.jp/~koba/
Brain NoiZ / こば

http://www.lofty-tec.co.jp/~koba/
Brain NoiZ / こば

http://www2s.biglobe.ne.jp/~crown/
CROWN / さくら

http://www-user.interq.or.jp/~naoki/
BIG BEN / N.I

http://www.tsp.ne.jp/~s_naoko/
Cafe pasta / 榊原薫奈緒子

http://www.tsp.ne.jp/~s_naoko/
Cafe pasta / 榊原薫奈緒子

http://www.tsp.ne.jp/~s_naoko/
Cafe pasta / 榊原薫奈緒子

http://www.tsp.ne.jp/~s_naoko/
Cafe pasta / 榊原薫奈緒子

http://www2k.biglobe.ne.jp/~candybox/
Candy Box / クリスト・マーク

http://www2k.biglobe.ne.jp/~candybox/
Candy Box / クリスト・マーク

http://www2.mwnet.or.jp/~tnaoko/
CHILDMOON / つくみや

http://www2.mwnet.or.jp/~tnaoko/
CHILDMOON / つくみや

http://www.din.or.jp/~sion/
Heaven's Gate / SION

利用網路介面可隨心所欲觀賞，而杏世也就是所謂的「BitmapResizer」free-ware（適用Windows95/NT4.0）這套軟體的作者，其可將原創圖片以不同比例來放大或縮小，真的很適合用在網路上。只要利用此特殊軟體，就能得到上述效果，而此軟體也可在同一網站上取得。

Flash animation effects are used throughout this site's interface in a very sophisticated way. Anze is also the creator of "Bitmap Resizer", a freeware program that uses an original algorithm for resizing images. This program is as good as or better than many high-end commercial graphic software and is a vailable for download from this site.

ORIGINAL

http://www2s.biglobe.ne.jp/~crown/
CROWN / さくら

http://www2.odn.ne.jp/~cae99980/
Deep / シン・オオバ

http://www.win.ne.jp/~doichan/
DOIchan! HomeStadium / DOIchan!

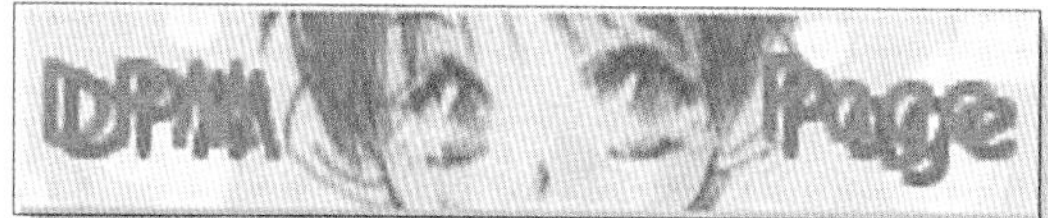

http://www.remus.dti.ne.jp/~dpm/
DPM-Page / 土方幹

http://www2e.biglobe.ne.jp/~earth/
EARTH CG Gallery / EARTH

http://www.mfi.or.jp/takehara/
FOREST GALLERY / 武原卓&しのぶ一樹

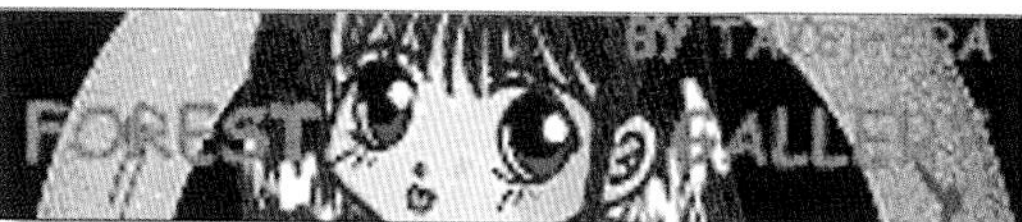

http://www.mfi.or.jp/takehara/
FOREST GALLERY / 武原卓&しのぶ一樹

http://www.mfi.or.jp/takehara/
FOREST GALLERY / 武原卓&しのぶ一樹

http://www.mfi.or.jp/takehara/
FOREST GALLERY / 武原卓&しのぶ一樹

http://www.na.rim.or.jp/~fre/
FRE's お絵かきページ / FRE

http://www.hh.iij4u.or.jp/~fun/
FUN / 不能花

http://www.threeweb.ad.jp/~bat6000/
GOLDEN BAT Home Page / GOLDEN BAT

http://www-user.interq.or.jp/~sashimi/index.html
グラタン帝国 / 真黒さしみ

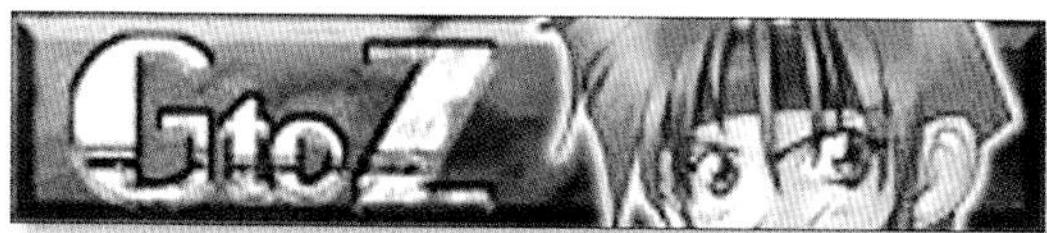

http://www.yui.or.jp/~gmetal/
G to Z / G-メタル

http://www.geocities.co.jp/SiliconValley/4718/
ICE's CYBER ROOM / 森永あいす

http://www.geocities.co.jp/SiliconValley/4718/
ICE's CYBER ROOM / 森永あいす

http://www.md.xaxon.ne.jp/~arcadia/
INNOCENT ARCADIA / 瑞輝 智佳

http://www.mars.dti.ne.jp/~kei/
Kei's Homepage / Kei

http://www.bekkoame.ne.jp/ro/gj13041/
小手川ゆあの極楽刑務所 / 小手川ゆあ

http://www2s.biglobe.ne.jp/~Marutaka/
CG Time / まるたか

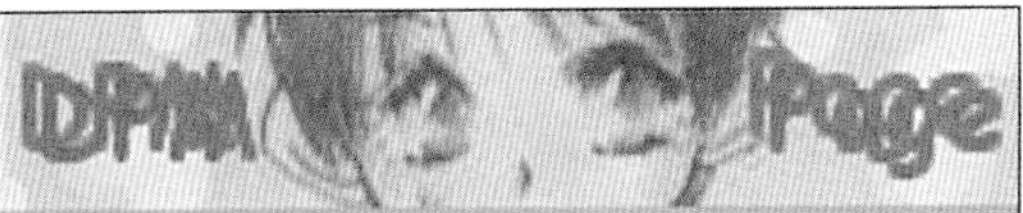

DPM-Page

http://www.remus.dti.ne.jp/~dpm/

土方幹

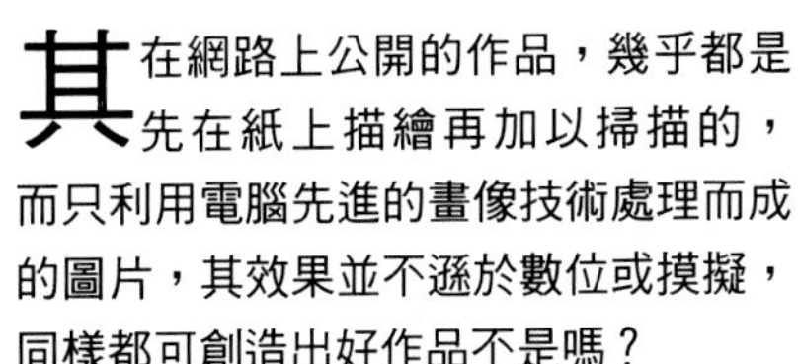

其在網路上公開的作品，幾乎都是先在紙上描繪再加以掃描的，而只利用電腦先進的畫像技術處理而成的圖片，其效果並不遜於數位或摸擬，同樣都可創造出好作品不是嗎？

Most of the images on this site are scans from hand-drawn works. Because of developments in the techniques of computer graphics software, however, you can't tell the difference between the scanned images and the digitally created images which are also shown here. In my opinion, both traditional and digital techniques are valuable and should be used as necessary to achieve the best results.

ORIGINAL

http://www.din.or.jp/~sanshiro/
SANSHIRO AKIYAMA'S Web page / 346

http://www.asahi-net.or.jp/~tr6h-kjur/
ぱんちゅ万歳 / h_k

http://www.asahi-net.or.jp/~tr6h-kjur/
ぱんちゅ万歳 / h_k

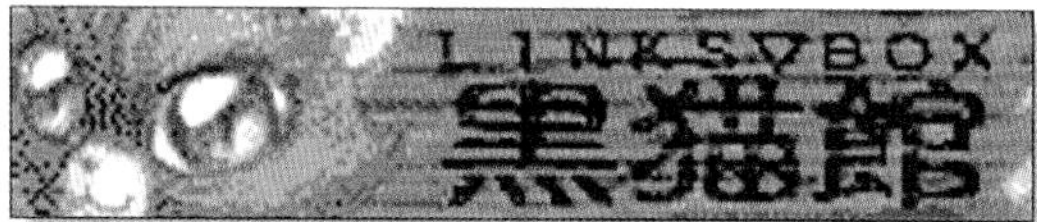

http://www.vega.or.jp/~shieo/
LINKS∇BOX ～黒猫館～ / 増多シイ夫

http://www.vega.or.jp/~shieo/
LINKS∇BOX ～黒猫館～ / 増多シイ夫

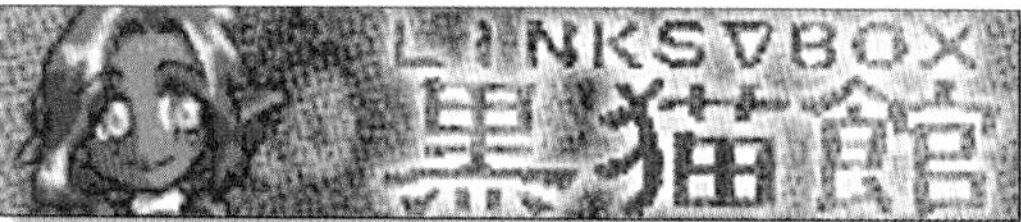

http://www.vega.or.jp/~shieo/
LINKS∇BOX ～黒猫館～ / 増多シイ夫

http://www.vega.or.jp/~shieo/
LINKS∇BOX ～黒猫館～ / 増多シイ夫

http://www2s.biglobe.ne.jp/~kaede_p/
めいぷる　りーふれっと～Kaede's CG Gallery / かえで☆

http://www.jah.ne.jp/~next96/
A RAILROAD JUNCTION / T.Kawasaki & YOU

http://www.jah.ne.jp/~next96/
A RAILROAD JUNCTION / T.Kawasaki & YOU

http://www.jah.ne.jp/~next96/
A RAILROAD JUNCTION / T.Kawasaki & YOU

http://www.jah.ne.jp/~next96/
A RAILROAD JUNCTION / T.Kawasaki & YOU

http://www.at-m.or.jp/~visnu/
すてすて～Stay Steady～ / とむそおや

http://www2q.biglobe.ne.jp/~SHIINA/
Rhapsody / K.Shiina

http://www2q.biglobe.ne.jp/~SHIINA/
Rhapsody / K.Shiina

http://www2q.biglobe.ne.jp/~SHIINA/
Rhapsody / K.Shiina

http://www.scan-net.or.jp/user/taidai/
Fragment Time / 泰大

http://www.din.or.jp/~sanshiro/
SANSHIRO AKIYAMA'S Web page / 346

すてすて～Stay Steady～
http://www.at-m.or.jp/~visnu/
とむそおや

作者我很喜愛這淡淡的色調以及「貓手」「貓尾巴」那濃密的感覺，雖然我也曾在其他作品中看到這種畫風，但主線部份與周圍色系相同，給人柔和的感覺。

I like the pastel colors and fluffy look of the cat paws and tail of this character. The use of matching outline colors and image colors in this and the other images here creates a nice soft look.

http://www.netlaputa.ne.jp/~naruse/
Karmic Relations! / 成瀬ちさと

http://www4.big.or.jp/~seneka/
NB学園 / 或十せねか

http://www4.big.or.jp/~seneka/
NB学園 / 或十せねか

http://www.phoenix-c.or.jp/~e-db8/
negative-doll / D.D

http://www.phoenix-c.or.jp/~e-db8/
negative-doll / D.D

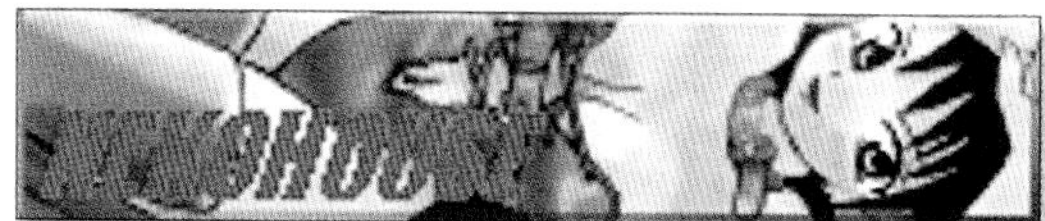

http://www2.tky.3web.ne.jp/~nekomimi/
NEKOHOUSE / 長居ねこ

http://www2.tky.3web.ne.jp/~nekomimi/
NEKOHOUSE / 長居ねこ

http://www2.tky.3web.ne.jp/~nekomimi/
NEKOHOUSE / 長居ねこ

http://www.pluto.dti.ne.jp/~winn/
NIGHT SHIFT / ういん

http://www.mirai.ne.jp/~niwaken/
NIWAKEN's HomePage / NIWAKEN

http://www.geocities.co.jp/Playtown-Denei/4194/
nobodyknows. / m_sugisaki

http://www.geocities.co.jp/Playtown-Denei/4194/
nobodyknows. / m_sugisaki

http://www.sam.hi-ho.ne.jp/~expo69/oko/
OKO HOME PAGE / oko

OKO HOME PAGE

http://www.sam.hi-ho.ne.jp/~expo69/oko/
OKO HOME PAGE / oko

http://www.seaple.icc.ne.jp/~dai/hp/
持続力-C

Karmic Relations!
http://www.netlaputa.ne.jp/~naruse/
成瀬ちさと

這是詳細闡述自己CG描繪手法的網頁，他似乎在解說著「這裏要怎麼畫」吧？，我個人很注意成瀨的色調調配，因為人物外形雖不討好，但「灰暗感」的配色卻是他個人極佳的風格。

More pages showing Chisato's CG graphic techniques will continue to be added to this site. She is also accepting requests so if you'd like to know how to do a particular technique, you should e-mail her. I especially like Chisato's choice of colors for her images. She uses a kind of muted color palette which gives her images a very special and original look.

Photo by (c)Tomoyuki.U http://www.yun.co.jp/~tomo/photo.html

http://www02.u-page.so-net.ne.jp/gb3/takizawa/
my confidence world / 上野真麻

http://www2.osk.3web.ne.jp/~gpxtaka/
Chocolate Paradise / GPX-TAKA

http://www2.osk.3web.ne.jp/~gpxtaka/
Chocolate Paradise / GPX-TAKA

http://www.sainet.or.jp/~cpunit/
CPU's CG Gallery / CPU

http://www.geocities.co.jp/Playtown/8756/
少年的電影箱 / 瑠堂れおな

http://www.din.or.jp/~sion/gen/
Ever Green Ever Blue / 源之助

http://www.ricoh.co.jp/src/people/fukuhara/
Phase Green / 川原由唯

http://www.ceres.dti.ne.jp/~snowdrop/
bit of GLASS FOREST / 里見桜次

http://www.kotonet.ne.jp/~tobo/
HAT home page Heavy DE Show!! / HAT

http://www.kotonet.ne.jp/~tobo/
HAT home page Heavy DE Show!! / HAT

http://www.infotrans.or.jp/~tabayan/
H&T Laboratory / 彪衣 亜騎（ひょうい あき）

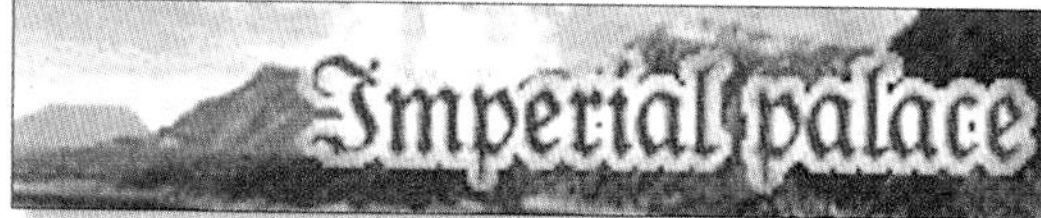

http://www.yo.rim.or.jp/~kwk36/
IMPERIAL PALACE / KWK36

http://www.sannet.ne.jp/userpage/ko-ji/
Memories / Coo

http://www.akira.ne.jp/~kasimu/(移転予定)
Welcome to the Rose Garden / 樫夢

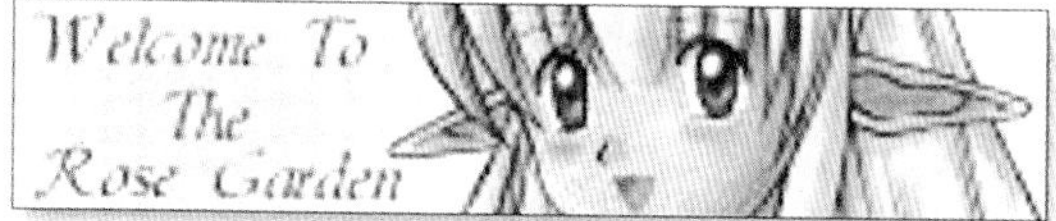

http://www.akira.ne.jp/~kasimu/(移転予定)
Welcome to the Rose Garden / 樫夢

http://www.sannet.ne.jp/userpage/ko-ji/
Memories / Coo

Welcome to the Rose Garden
http://www.akira.ne.jp/~kasimu/へ移転予定
樫夢

在選擇網站的操作指令就如同在玩探險遊戲一般，而要如何連結到你所需要的網站呢？網路精靈-露菲會教你許多上網的樂趣，依據存取時間可顯示不同的網站，這樣的組合也很不錯喔-

The design of this site is like a point and click adventure game interface. When you select the links, a fairy named "Rufy" gives you special guidance. We already have sites that do such things as change according to the time you visit but the interface for this site is quite unique and the elements work well together.

http://www.kotonet.ne.jp/~tobo/
HAT home page Heavy DE Show!! / HAT

http://www.osk.3web.ne.jp/~catslabo/
Cat's Labo / ねこやま工房

http://www.biwa.ne.jp/~k-t/
KIYO'S CG ISLAND / KIYO

http://www.sannet.ne.jp/userpage/leon/main.html
LEONPURPLE CG WORLD / れおんぱあぷる

http://www.geocities.co.jp/Playtown/3331/
LEVEL / NaoXYZ

http://www.remus.dti.ne.jp/~ryouma/
Lincoln Island / 竜馬

http://www.remus.dti.ne.jp/~ryouma/
Lincoln Island / 竜馬

http://www1.odn.ne.jp/ishida/
Life Like Love / 石田あきら

http://www.osk.3web.ne.jp/~ma2an/
Squid Trap / まっつあん

http://www.osk.3web.ne.jp/~ma2an/
Squid Trap / まっつあん

sana's
phantasmagoria

http://www2s.biglobe.ne.jp/~SANA/
Phantasmagoria / 紗那

http://www2s.biglobe.ne.jp/~SANA/
Phantasmagoria / 紗那

http://www2s.biglobe.ne.jp/~SANA/
Phantasmagoria / 紗那

http://www.din.or.jp/~tugiri/
T.O.H.P. / 津霧みん

http://www.din.or.jp/~tugiri/
T.O.H.P. / 津霧みん

Phantasmagoria
http://www2s.biglobe.ne.jp/~SANA/
紗那

時尚設計師一紗那所描繪的作品，頗具時尚色彩，果然深具這種特色的作品，不僅能創造出利潤，同時更能創造別出心裁的服飾。

The clothing designs of fashion designer Sana's characters are very charming. Her talent for fashion design is obviously a big plus in creating these images. The clothing she designs for her characters is also designed to be actually made and worn.

http://www.rnac.or.jp/~akeo/
MONA'S MONA / akeo

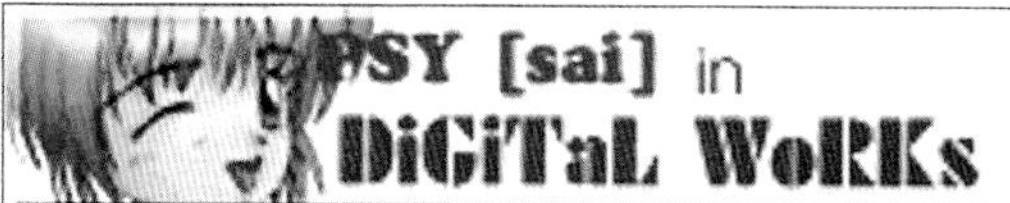

http://www.os.rim.or.jp/~psy/
PSY [sai] in DiSiTaL WoRKs / PSY [sai]

http://www02.so-net.ne.jp/~quattro/
quattro Hobbs Room / くわとろ

http://www2s.biglobe.ne.jp/~n_kato/
Mirage Field / レッド

http://www.os.rim.or.jp/~report/
REPORT / 日高司

http://home2.highway.ne.jp/socket5/
ろりろりメルトダウン / 左向き矢印

http://www.angel.ne.jp/~ryoji/
EDEN the DOORS official page / 氷堂涼二

http://www.angel.ne.jp/~ryoji/
EDEN the DOORS official page / 氷堂涼二

http://www.din.or.jp/~rinda/
Ring Ring Mix! / りんだ

http://www.din.or.jp/~rinda/
Ring Ring Mix! / りんだ

http://www.din.or.jp/~rinda/
Ring Ring Mix! / りんだ

http://www.din.or.jp/~rinda/
Ring Ring Mix! / りんだ

http://www.yo.rim.or.jp/~sapara/
Saparak / さぱら

http://www2s.biglobe.ne.jp/~satox/
SATOXのホ～ムペ～ジ/ SATOX

http://www1.odn.ne.jp/kaede/
時雨堂 / 水月楓

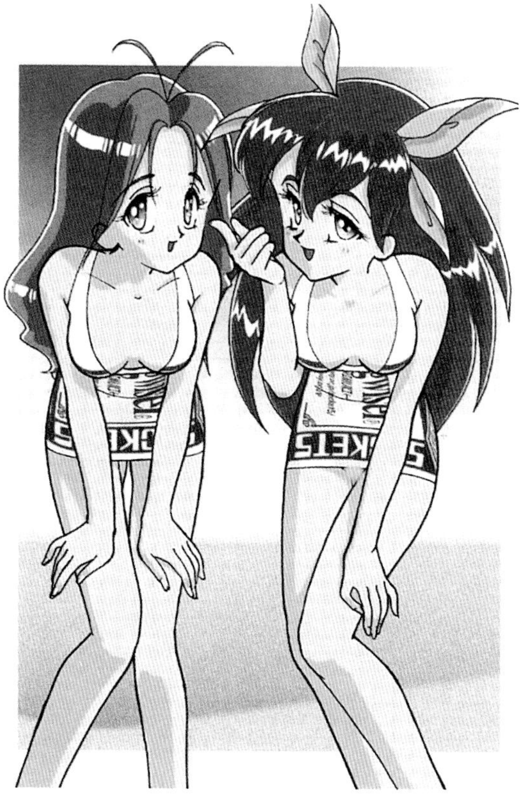

http://home2.highway.ne.jp/socket5/
ろりろりメルトダウン / 左向き矢印

Ring Ring Mix!

http://www.din.or.jp/~rinda/

りんだ

RINDA主辦的「Painter-ml」，是為了統合畫家製作成冊，實際上我也受到邀請。利用便宜且功能不錯的電腦，就能作出如此棒的作品，這真是個不錯的時代。

Rinda runs a mailing list for manga drawing and CG techniques called "Painter-ml". I am also a member of this mailing list. Inexpensive high quality computers and CG mailing lists -- what a great age to be living in!

http://www.alles.or.jp/~s2r/
SALTSHAKER / 塩原信一

http://www2.tky.3web.ne.jp/~syunichi/
SEVEN FLAVOR SPICE / しゅんにい

http://www.din.or.jp/~ko-cha/
TEA・ROOM KO-CHA's Web Page / こ～ちゃ

http://www1.pos.to/~takuma/
TARAKO FACTORY / たくま朋正

http://www.lares.dti.ne.jp/~youma/coterie/
YouMa in Coterie / ゆうまじろう

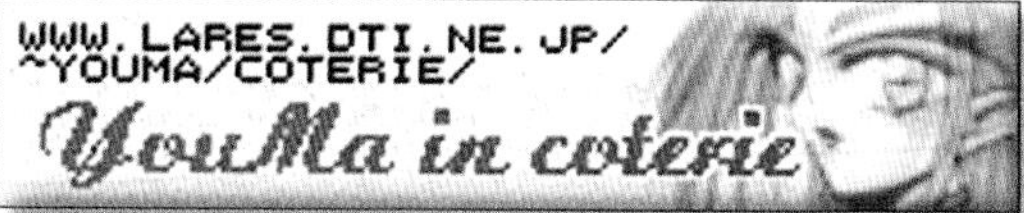

http://www.lares.dti.ne.jp/~youma/coterie/
YouMa in Coterie / ゆうまじろう

http://www.lares.dti.ne.jp/~youma/coterie/
YouMa in Coterie / ゆうまじろう

http://www.lares.dti.ne.jp/~youma/coterie/
YouMa in Coterie / ゆうまじろう

http://www.lares.dti.ne.jp/~youma/coterie/
YouMa in Coterie/ ゆうまじろう

http://www.lares.dti.ne.jp/~youma/coterie/
YouMa in Coterie/ ゆうまじろう

http://www.lares.dti.ne.jp/~youma/coterie/
YouMa in Coterie / ゆうまじろう

http://www.lares.dti.ne.jp/~youma/coterie/
YouMa in Coterie / ゆうまじろう

http://www.lares.dti.ne.jp/~youma/coterie/
YouMa in Coterie / ゆうまじろう

http://www.lares.dti.ne.jp/~youma/coterie/
YouMa in Coterie / ゆうまじろう

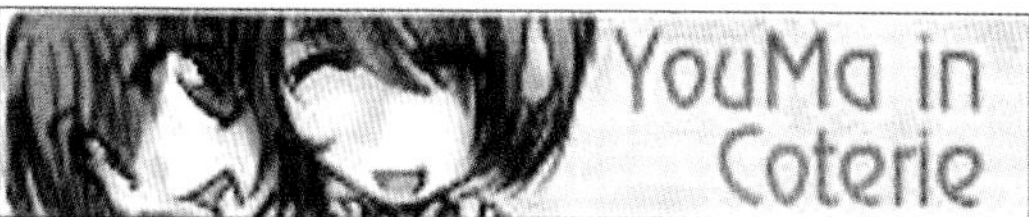

http://www.lares.dti.ne.jp/~youma/coterie/
YouMa in Coterie / ゆうまじろう

http://www.lares.dti.ne.jp/~youma/coterie/
YouMa in Coterie / ゆうまじろう

http://www.lares.dti.ne.jp/~youma/coterie/
YouMa in Coterie / ゆうまじろう

http://www.venus.dti.ne.jp/~uniuni/
UniUni's HomePage / うにうに

http://www.venus.dti.ne.jp/~uniuni/
UniUni's HomePage / うにうに

http://www.vc-net.or.jp/~vishiki/
VALKYRIE ILLUSION / 士貴貴ゆう

SALTSHAKER

Ring Ring Mix!
http://www.din.or.jp/~rinda/
りんだ

鹽原的網站中有著頂級的網頁，雖然其網頁打著「手法乾淨俐落－由鉛筆、原子筆、麥克筆所組成，相當簡單的作品」的宣傳標語，但實際上卻是很不簡單，那是融合幻想的未來世界的世界性作品。

I like the catchphrase that opens this site, "light salty taste -- a cheap flavor composed of pencils, ball point pens and the Mac". In reality though, there is nothing cheap-feeling at all about this site. The images you can see here are a combination of fantasy themes and the world of the near future.

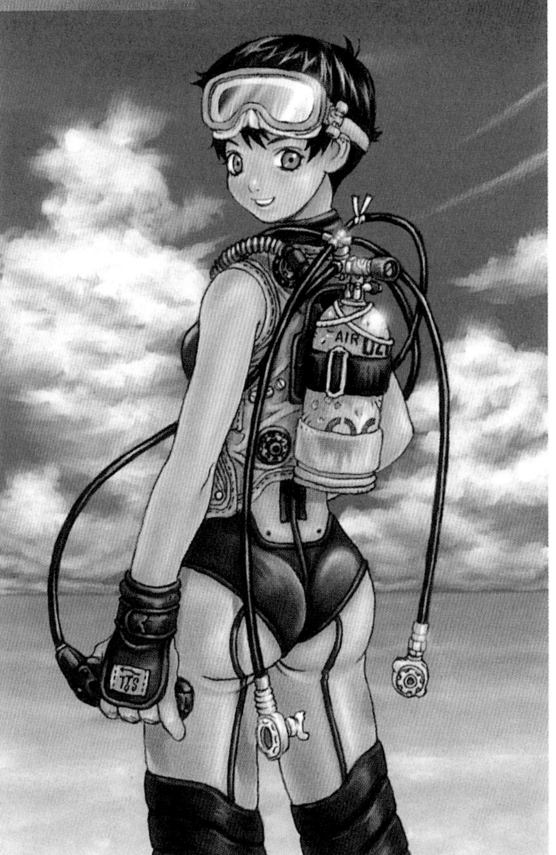

ORIGINAL

http://www7.peanet.ne.jp/~dai_nn/
Scrap machines /大王

http://www01.u-page.so-net.ne.jp/db3/fifth/
Page5 / FIFTH

http://www01.u-page.so-net.ne.jp/db3/fifth/
Page5 / FIFTH

http://www01.u-page.so-net.ne.jp/db3/fifth/
Page5 / FIFTH

http://www2m.biglobe.ne.jp/~mizuha/
IMPLACABLE / 水羽輝幸

http://www2m.biglobe.ne.jp/~mizuha/
IMPLACABLE / 水羽輝幸

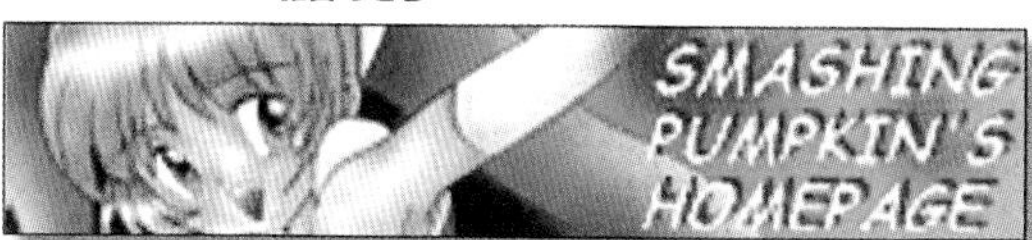

http://www.os.rim.or.jp/~miyabie/
MILKY WATER / 雅日あきら

http://www.fuji.ne.jp/~sumner/
SMASHING PUMPKIN'S HOMEPAGE / さむな

http://www.studio-zero.com/
STUDIO ZERO / JACK!zero

http://www.studio-zero.com/
STUDIO ZERO / JACK!zero

http://member.nifty.ne.jp/suishoudo/
atelier-Suite Grove- / 睡宵堂

http://www.imasy.or.jp/~wakachan/
wakachan's HomePage / 和佳-chan

http://www.imasy.or.jp/~wakachan/
wakachan's HomePage / 和佳-chan

http://wakachan.kaynet.or.jp/
wakachan's HomePage / 和佳-chan

http://member.nifty.ne.jp/suishoudo/
atelier-Suite Grove- / 睡宵堂

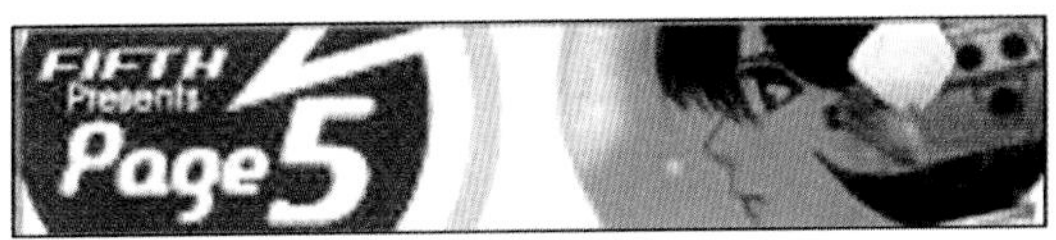

Page5
http://www01.u-page.so-net.ne.jp/db3/fifth/
FIFTH

從FIFTH宛如描繪於紙上般的筆觸，讓人感覺溫暖，網站中還登載著其他相當奇特的作品。

Here are some selections from Fifth's warm pen and ink style graphics. Fifth also creates fantasy and SF style works.

http://www1.sphere.ne.jp/emcweb/megababes/
megababes / emico and kilkeny

http://www.win.ne.jp/~sonson/
sonson's Cafe / そんそん

http://www.pluto.dti.ne.jp/~seiya-n/
StOneBOX（すと〜んぼっくす）/ 野沢醒矢

http://www.pluto.dti.ne.jp/~seiya-n/
StOneBOX（すと〜んぼっくす）/ 野沢醒矢

http://member.nifty.ne.jp/CATSFACE/
Cat's Face / 水花

http://member.nifty.ne.jp/CATSFACE/
Cat's Face / 水花

http://member.nifty.ne.jp/CATSFACE/
Cat's Face / 水花

http://member.nifty.ne.jp/CATSFACE/
Cat's Face / 水花

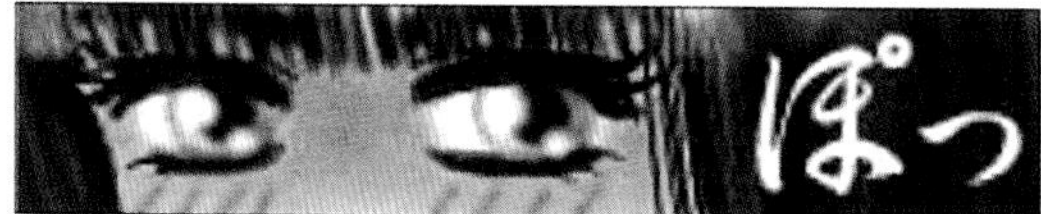

http://member.nifty.ne.jp/CATSFACE/
Cat's Face / 水花

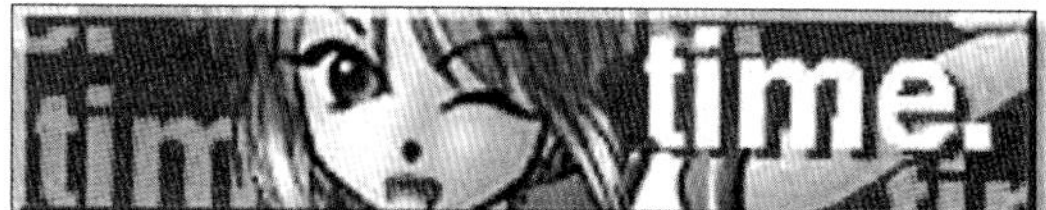

http://www.geocities.co.jp/Playtown/4029/
time. / きずき ゆずる

http://www.geocities.co.jp/Playtown/4029/
time. / きずき ゆずる

http://www.geocities.co.jp/Playtown/4029/
time. / きずき ゆずる

http://www.geocities.co.jp/Playtown/4029/
time. / きずき ゆずる

http://www.geocities.co.jp/Playtown/4029/
time. / きずき ゆずる

http://www1.sphere.ne.jp/emcweb/megababes/
megababes / emico and kilkeny

Cat's Face
http://member.nifty.ne.jp/CATSFACE/
水花

公開這些以描繪「夏天」「天空」「星星」「花朵」為主題的作品，在所有網站中都洋溢著貓耳風情的作品，另外此網站還看得見以天使為題材的作品。

This site has images with themes relating to summer, the sky, stars and flowers. Most of the characters depicted here are cat-eared types but there are also soft focus style images of angels characters displayed.

http://www.geocities.co.jp/SiliconValleyPaloAlto/1315/kai_index.html
UNSTABLEST DESIGNxxx / カイ

http://www.sainet.or.jp/~itokei/
Wonder-Ranch / いとけい

http://www.pluto.dti.ne.jp/~hitom/
JAPOTIO ATOLI / 人身

http://www.pluto.dti.ne.jp/~hitom/
JAPOTIO ATOLI/ 人身

http://www.t3.rim.or.jp/~kairi/
Open Your Eyes / 海里

http://www.t3.rim.or.jp/~kairi/
Open Your Eyes / 海里

http://www.t3.rim.or.jp/~kairi/
Open Your Eyes / 海里

http://www.nerimadors.or.jp/~maguro/
MAGUROMEDIA / まぐろせんせい

http://miya.or.jp/~mochiex/
もちもち本舗 / EXP

http://www1.plala.or.jp/milk/
MILK COCOA CAN / 欧柊ここあ

http://www.geocities.co.jp/Playtown/5676/
ミノンのお部屋 / NON

http://www.linkclub.or.jp/~minakami/
MarineMarilyn HomePage / みなかみ

http://www.din.or.jp/~otimusha/
Knight of Dusk / 落ち武者

http://www.venus.dti.ne.jp/~nhodzmi/
Vertigo High / ほづみないき

http://www.din.or.jp/~otimusha/
Knight of Dusk / 落ち武者

Vertigo High

http://www.venus.dti.ne.jp/~nhodzmi/

ほづみないき

描繪同一人作品，充滿了作者個人風格的CG作品，HOTSUMI描繪的女孩散發著莫名的悲傷的神情，給人寧靜的印象。

"Doujin" or fanzine characters and also original CG characters are the focus of this site. Naiki's girl characters have a kind of sad wistful look. To me they also seem very feminine.

http://www.sun-inet.or.jp/~sanpei/
TRAIN OF THOUGHT / 3PAY HIRO

http://www2a.biglobe.ne.jp/~acharin/
ACHARIN's WORLD / ACHARIN

http://www2s.biglobe.ne.jp/~AOKINAO/
あおきなおウェブページ / あおきなお

http://www.elf.ne.jp/~aki/
Downstairs Sweetheart - 階下の恋人 - / さくらんぼ

http://www.elf.ne.jp/~aki/
Downstairs Sweetheart - 階下の恋人 - / さくらんぼ

http://www.elf.ne.jp/~aki/
Downstairs Sweetheart - 階下の恋人 - / さくらんぼ

http://www.elf.ne.jp/~aki/
Downstairs Sweetheart - 階下の恋人 - / さくらんぼ

http://www.elf.ne.jp/~aki/
Downstairs Sweetheart - 階下の恋人 - / さくらんぼ

http://www.elf.ne.jp/~aki/
Downstairs Sweetheart - 階下の恋人 - / さくらんぼ

http://www.elf.ne.jp/~aki/
Downstairs Sweetheart - 階下の恋人 - / さくらんぼ

http://www.sanynet.ne.jp/~nori/
My Taste / Nori

http://www4.big.or.jp/~azuna/pinkgun/
PINKGUN / あずなしあ

http://www4.big.or.jp/~azuna/pinkgun/
PINKGUN / あずなしあ

http://www.angel.ne.jp/~gazer/you/index.htm
Olive Free State / 高槻 悠

Downstairs Sweetheart - 階下の恋人 -

http://www.elf.ne.jp/~aki/

さくらんぼ

PUNI中的PUNI，作品中的小薰（登載於此的CG角色）很是可愛，隨著季節的變化，作者我也期待著小薰的蛻變。

This home page has the best of the "puni" style characters. Kaori-chan, shown here, has very cutely drawn lips and cheeks. I look forward to seeing Kaori-chan's different looks as they change with the season.

http://www2s.biglobe.ne.jp/~kas/
Negative X / kasun

http://www.tk.xaxon.ne.jp/~mao_h/
DOUBLE TAP / 榛名まお

http://www2s.biglobe.ne.jp/~fromy/
FROM 雪待月 / 佐久間 和弥

http://www2s.biglobe.ne.jp/~fromy/
FROM 雪待月 / 佐久間 和弥

http://plaza3.mbn.or.jp/~mugen_hikoh/
夢幻飛行 / 瑞枝 悠

http://www.lares.dti.ne.jp/~hazeou/
KARASAGI / 立花はぜお

http://www.lares.dti.ne.jp/~hazeou/
KARASAGI / 立花はぜお

http://www.lares.dti.ne.jp/~hazeou/
KARASAGI / 立花はぜお

http://www.lares.dti.ne.jp/~hazeou/
KARASAGI / 立花はぜお

http://www.lares.dti.ne.jp/~hazeou/
KARASAGI / 立花はぜお

http://home2.highway.or.jp/mmoriya/mieko/
MIEKO'S PAGE / みえこ

http://home2.highway.or.jp/mmoriya/mieko/
MIEKO'S PAGE / みえこ

http://nekoneko.com/nekotokoworks/
ねことこWorks / ねこねこ＆とことこ

http://nekoneko.com/nekotokoworks/
ねことこWorks / ねこねこ＆とことこ

http://nekoneko.com/nekotokoworks/
ねことこWorks / ねこねこ＆とことこ

http://www.tk.xaxon.ne.jp/~mao_h/
DOUBLE TAP / 榛名まお

ねことこWorks
http://nekoneko.com/nekotokoworks/
ねこねこ&とことこ

這是由貓小姐和TOKOTOKO兩位所組成的網站，就如同他們的名字一般，這是洋溢著滿滿「貓耳」特色的CG，我在想看過之後，忍不著愛上了它，這讓我更加感謝更紗了。

This web site is run by the the "circle" of Neko Neko and Toko Toko. As you might guess from Neko Neko's nickname this site is full of "neko"-eared cat girl characters. I was also very pleased to find images of one of my favorite manga characters here.

ORIGINAL

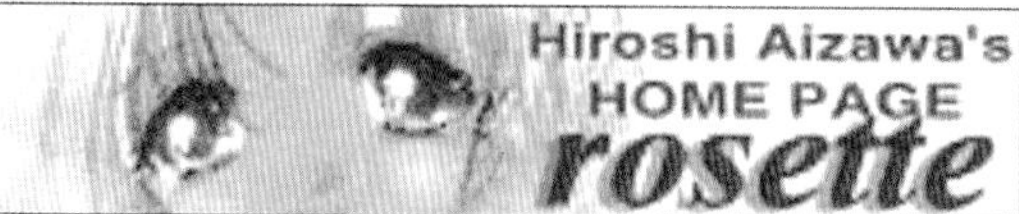

http://www02.so-net.ne.jp/~highrisk/proom/aizawatop.html
rosette / あいざわひろし

http://www.nsknet.or.jp/~akuta/
2B鉛筆工房 / 芥

http://www.nsknet.or.jp/~akuta/
2B鉛筆工房 / 芥

http://www.nsknet.or.jp/~akuta/
2B鉛筆工房 / 芥

http://www.win.ne.jp/~non/
It's "NON" Parity!! / NON

http://www.os.rim.or.jp/~charm/
CHARMING SQUARE GARDEN / CHARM

http://www.linkclub.or.jp/~shinno/
Jasmine Tea Break / Jasmine Tea

http://www.ceres.dti.ne.jp/~iwadate/
Rock Climbing / 岩舘こう

http://www.mars.dti.ne.jp/~kuzuhara/
Konomi Kuzuhara's Homepage / 葛原このみ

http://www.tt.rim.or.jp/~sphere/
Pastel Image / SPHERE

http://www.tt.rim.or.jp/~sphere/
Pastel Image / SPHERE

http://www.tt.rim.or.jp/~tagro/
TARGO's WorkShop / TARGO

http://www.scarecrow.co.jp/~yabou/
ケン太+樹のCG「野望のページ」/ ケン太&樹

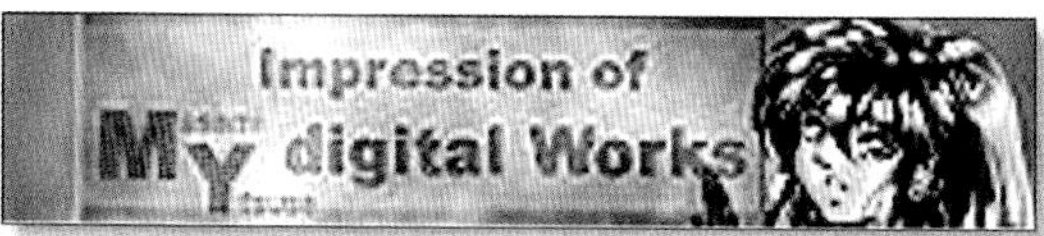

http://www.din.or.jp/~yitsuse/
Impression of MY digital works / 愛瀬雅巳

http://www2s.biglobe.ne.jp/~ym_page/art/
WHITE CREATION / 増田幸紀

http://www02.so-net.ne.jp/~y_co/
TECHNO LIFE / 松野唯

http://www02.so-net.ne.jp/~y_co/
TECHNO LIFE / 松野唯

http://www02.so-net.ne.jp/~y_co/
TECHNO LIFE / 松野唯

http://www02.so-net.ne.jp/~y_co/
TECHNO LIFE / 松野唯

Rock Climbing

http://www.ceres.dti.ne.jp/~iwadate/

岩舘こう

因為岩館提供了色彩非常鮮明的作品，而印刷則只為了希望能展現出其完美的風格・・。

言館除了原創作品之外，還有許多以遊戲為題材的作品，每一個女孩的臉上都散發著岩館的風格。

Kou contributed this image with the vivid blue color. I just wish the blue in this printed version was as bright as in the original... In addition to original images there are many images of scenes from games. I think the innocent-looking faces of Kou's characters are very cute.

ORIGINAL

http://www2.gol.com/users/batcave/
A.S.PANORAMIC / 佐藤明機

http://www2.gol.com/users/batcave/
A.S.PANORAMIC / 佐藤明機

http://www2.gol.com/users/batcave/
A.S.PANORAMIC / 佐藤明機

http://www2.gol.com/users/batcave/
A.S.PANORAMIC / 佐藤明機

http://home.interlink.or.jp/~akituki/
秋月書店 / 秋月つき

http://www.netlaputa.ne.jp/~dap/
DAP- ENTERPRISE / だっぷ

http://www.netlaputa.ne.jp/~dap/
DAP- ENTERPRISE / だっぷ

http://www.bekkoame.ne.jp/ro/yodomi/
溝川淀美の部屋 / 溝川淀美

http://www.annie.ne.jp/~sat/
D.P.I. / Sat

http://www.alles.or.jp/~knot/
Eye Appeal / 結稀

http://www.ddt.or.jp/~naka/
ILLUST-GUERRILLA / NAKA

http://www.ddt.or.jp/~naka/
ILLUST-GUERRILLA / NAKA

http://www.ddt.or.jp/~naka/
ILLUST-GUERRILLA / NAKA

http://www.ddt.or.jp/~naka/
ILLUST-GUERRILLA / NAKA

http://www.kt.rim.or.jp/~hymz/
Delusions of grandeus / MON

http://www.asahi-net.or.jp/~rd8t-mtmt/
SHO ROOM / 松本彰

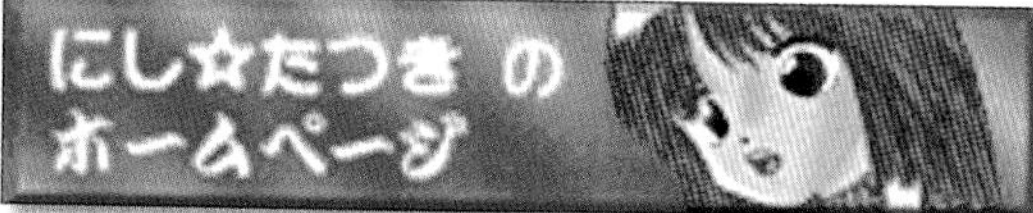

http://www.asahi-net.or.jp/~kd4n-nsmt/stardust.htm
NishiTatsuki's Home Page! / にし☆たつき

http://www.asahi-net.or.jp/~kd4n-nsmt/stardust.htm
NishiTatsuki's Home Page! / にし☆たつき

http://www02.so-net.ne.jp/~wolf/
ときめき＿BOX / 狼太郎

ときめき＿BOX
http://www02.so-net.ne.jp/~wolf/
狼太郎

狼太郎身為一位職業漫畫家，不論是在雜誌上或連環漫畫領域都相當地活躍，網站上也詳細地解說他的CG的描繪手法，另外還設置問題屋以解開網客心中的疑問。

The author of this page is a professional manga illustrator who contributes to various magazines and comic series. His web site has a detailed "how to" section on creating CG illustrations. There is also a message board to which you can post requests for "how to's" on techniques you would like to learn.

http://www02.so-net.ne.jp/~asyura/
あしゅらのおもちゃ箱・はいぱ～！ / 雅あしゅら

http://www.baba-t.com/~cotoba/
ことばの国 / COTOBA

http://www.capricorn.cse.kyutech.ac.jp/~ei/UC/
えいくんち / えいくん

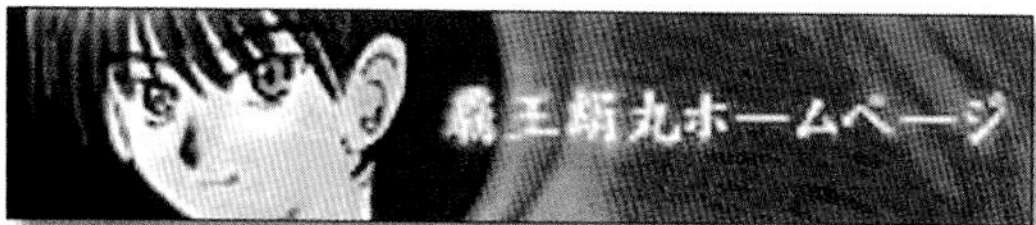

http://www.future.ne.jp/kinumaru/
Twinkle Town / 覇王絹丸

http://www.future.ne.jp/kinumaru/
Twinkle Town / 覇王絹丸

http://cgi3.osk.3web.ne.jp/~mkty/
P.H.S.(ぷに・ほえ・ステーション) / かかか

http://cgi3.osk.3web.ne.jp/~mkty/
P.H.S.(ぷに・ほえ・ステーション) / かかか

http://www.ceres.dti.ne.jp/~omecha/
藤岡建設 / 藤岡建機

http://www.cyberoz.net/city/kohi/index.html
ふつーのほーむぺーじ / こうひい

http://www.din.or.jp/~kotobuki/
FORZA on the Web / 寿圭祐

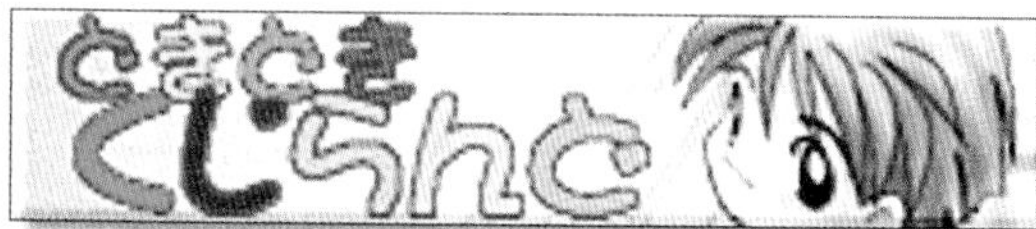

http://www.cc.rim.or.jp/~hinobe/
どきどきくじらんど / 火延 真

http://www.cc.rim.or.jp/~hinobe/
どきどきくじらんど / 火延 真

http://www.cc.rim.or.jp/~hinobe/
どきどきくじらんど / 火延 真

http://www.cc.rim.or.jp/~hinobe/
どきどきくじらんど / 火延 真

http://www.kt.rim.or.jp/~myu2/
Studio Myu. / 山ちゃん

http://www.kt.rim.or.jp/~myu2/
Studio Myu. / 山ちゃん

http://www.angel.ne.jp/~nochina/
よこむきもぐら / ハムくまズ

http://www.asaka.ne.jp/~masaru/
夢色鉛筆 / まさる

P. H. S.（ぷに・ほえ・ステーション）

http://cgi3.osk.3web.ne.jp/~mkty/

かかか

登載著許多神采飛揚女子的CG作品，那是KAKAKA的網站，如今已是利用電腦，來處理彩色圖片的時代了，由於專業技術的普及，16色及全色都能融入在CG作品中，而我覺得現在的16色或256色的CG，已呈現出不能再好的作品了。

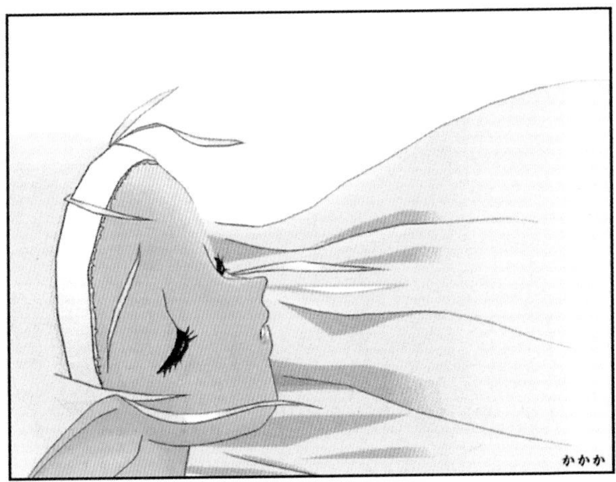

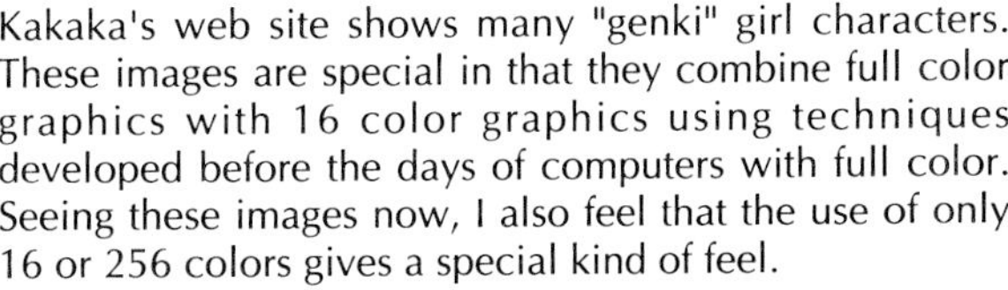

Kakaka's web site shows many "genki" girl characters. These images are special in that they combine full color graphics with 16 color graphics using techniques developed before the days of computers with full color. Seeing these images now, I also feel that the use of only 16 or 256 colors gives a special kind of feel.

http://member.nifty.ne.jp/yasumi/
第三新東京星　天使居住区 / 嵩石恭巳

http://member.nifty.ne.jp/yasumi/
第三新東京星　天使居住区 / 嵩石恭巳

http://member.nifty.ne.jp/BADCOMPANY/
BADCOMPANY / 恋川春町

http://home2.highway.or.jp/naono/
Persian Pudding / 天野 忍

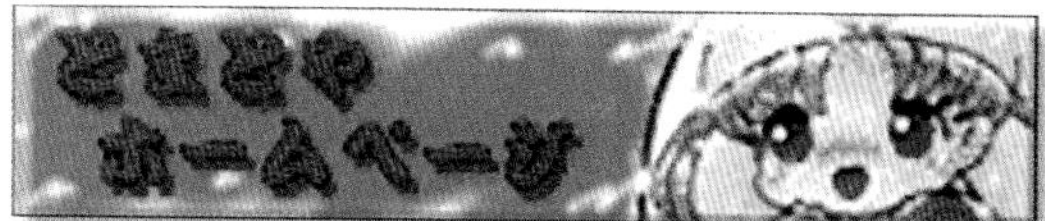

http://www.aaa-int.or.jp/tomato/
とまとやホームページ / 冷やしたとまと

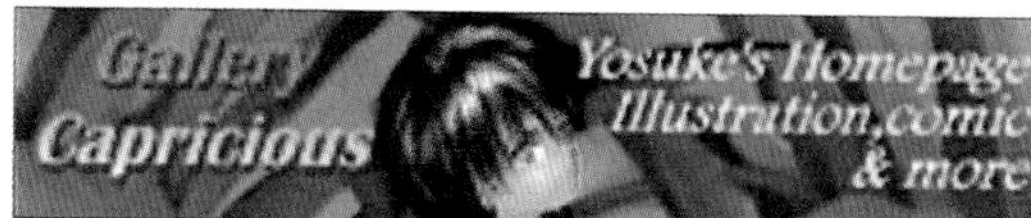

http://plaza7.mbn.or.jp/~yosukef/
Gallery Capricious / よーすけ

http://plaza7.mbn.or.jp/~yosukef/
Gallery Capricious / よーすけ

http://plaza7.mbn.or.jp/~yosukef/
Gallery Capricious / よーすけ

http://plaza7.mbn.or.jp/~yosukef/
Gallery Capricious / よーすけ

http://plaza7.mbn.or.jp/~yosukef/
Gallery Capricious / よーすけ

http://plaza7.mbn.or.jp/~yosukef/
Gallery Capricious / よーすけ

http://plaza7.mbn.or.jp/~yosukef/
Gallery Capricious / よーすけ

http://www.asahi-net.or.jp/~SJ8T-YSMT/
風雀のHP / 風雀

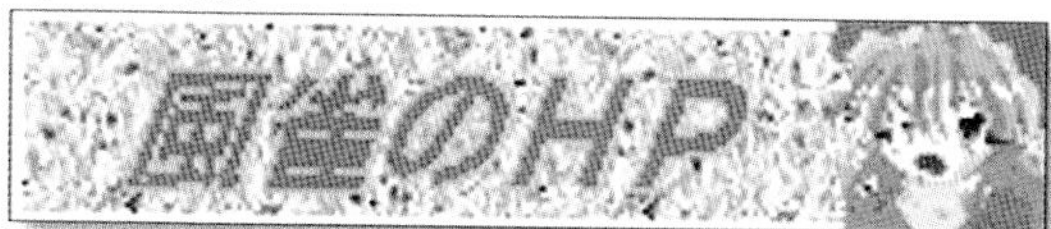

http://www.asahi-net.or.jp/~SJ8T-YSMT/
風雀のHP / 風雀

http://member.nifty.ne.jp/pbm/
PARAIBA TOURMARINE / 石田真由美

http://member.nifty.ne.jp/pbm/
PARAIBA TOURMARINE / 石田真由美

http://home.att.ne.jp/red/miz/
篝火工房 / 水谷ヨージ

http://www.alles.or.jp/pub/nezupon/
PON'S-BRAND / 大乃国ぽん

http://w32.mtci.ne.jp/~rito/
PURPLE CASTLE -in MATSUKAZE Pub.- / 麺次

http://member.nifty.ne.jp/tsunoda/
ディラバル ブラウザ / 代由

Gallery Capricious

http://plaza7.mbn.or.jp/~yosukef/

よーすけ

看到描繪得栩栩如生的綠色熱帶植物以及少女的作品，讓我非常地感動。他完全展現出描繪在紙上所沒辦法展現出的色彩鮮明度，另外還有以卡通或漫畫為題材的作品。

I was impressed by this image of such a lively energetic girl surrounded by very vivid green tropical plants. Because this printed version can't truly capture the brightness of the green in the original CG image, I highly recommend that you visit this site to take a look at the original and also the artist's many other images of anime and game characters.

http://plaza4.mbn.or.jp/~kuronekokan/
黒猫館 / 篠崎広里

http://www.alles.or.jp/~gin/INDEX.HTM
銀龍牙'sほうむぺえじ / 銀龍牙

http://hot.netizen.or.jp/~akari/
橋本組出張課託児所 / 兼越ゆうじ

http://plaza28.mbn.or.jp/~inukai/jinjya/
犬飼神社 / 真田丸マサオ

http://www.big.or.jp/~hannaya/
J.SAJI'S Homepage / 佐治ジロー

http://home3.highway.ne.jp/matsuri/
K-いんぱくと / 松莉 詠

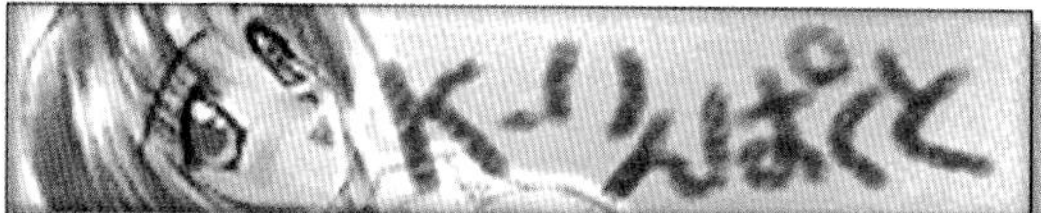

http://home3.highway.ne.jp/matsuri/
K-いんぱくと / 松莉 詠 / 松莉 詠

http://www.alles.or.jp/~walkure/
三つ編み萌え / 紫 らない

http://member.nifty.ne.jp/miyanon/
絢爛漫画遊戯館宮野式 / 宮野あみか

http://member.nifty.ne.jp/morii/
狂乱麗舞 /光明公

http://plaza8.mbn.or.jp/~masaoki/
Masaoki Satou's Homepage / 佐藤正興

http://home.interlink.or.jp/~okazakir/
One From The Heart / 剣崎 星紅

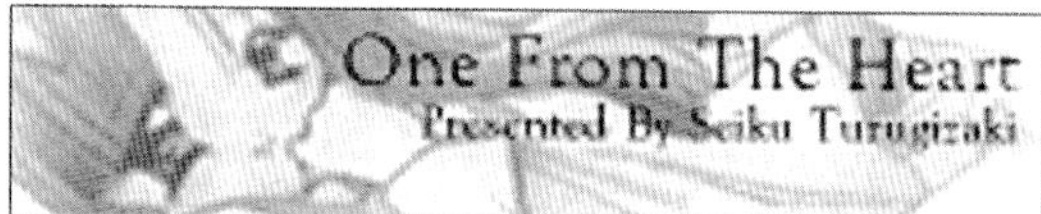

http://home.interlink.or.jp/~okazakir/
One From The Heart / 剣崎 星紅

http://www.alles.or.jp/~tomos/
パイナップル畑 / 桜餅 智

http://members.aol.com/SASAMIC43/
はあとぎゃらりー / 今井一成

http://csefs01.ce.nihon-u.ac.jp/~u086077/
Seraphic / 砂月雪人

http://www.alles.or.jp/~karin/
Sion's Web Page / 紫音

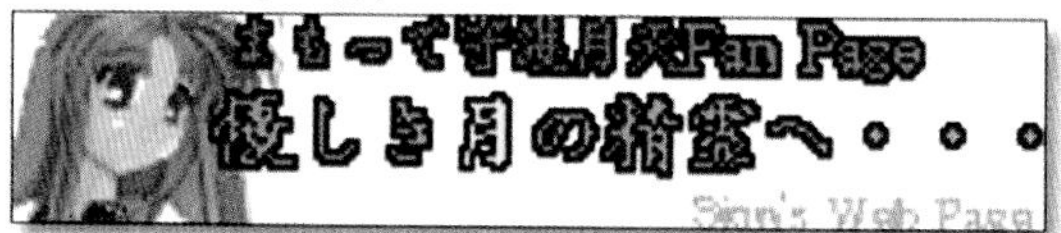

http://www.alles.or.jp/~karin/
Sion's Web Page / 紫音

http://plaza29.mbn.or.jp/~SAKURA/
SCHIZOPHOTIC / 佐倉修造

http://deniam.com/users/xeronion/
xeronion's work / xeronion

xeronion's work
http://deniam.com/users/xeronion/
xeronion

描繪這個火精靈的是韓國的ZELOMION，希望傳達的是韓國也有許多精通於繪畫的人才，而我也是這麼認為，這是以登載日本動畫及遊戲為主要題材的網站。

This image of a person controlling fire spirits is by Xeronian who comes from Korea. Xeronian wants everyone to know that Korea also has many fine artists. I agree. This web site presents original characters and also Japanese anime and game characters.

http://s3.hopemoon.com/~elx001/
備前屋 / ひの

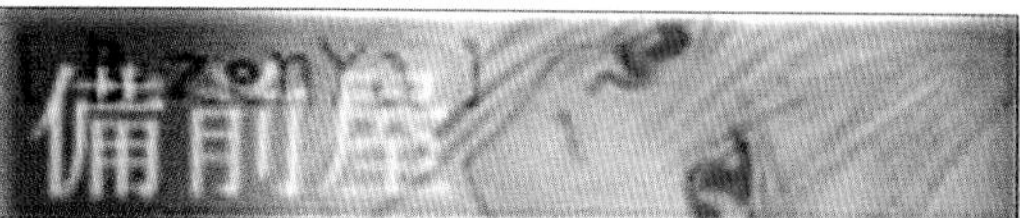

http://s3.hopemoon.com/~elx001/
備前屋 / ひの

http://s3.hopemoon.com/~elx001/
備前屋 / ひの

http://s3.hopemoon.com/~elx001/
備前屋 / ひの

http://s3.hopemoon.com/~elx001/
備前屋 / ひの

http://www.bekkoame.ne.jp/ha/mikage/
零の光景 / みかげ

http://www.asahi-net.or.jp/~MN2Y-TBT/
yugori page / yugori

http://plaza15.mbn.or.jp/~norino/
のりのホームページ / のり

http://www.din.or.jp/~acht/index.html
LAZY VIRUS / たにざきひなり

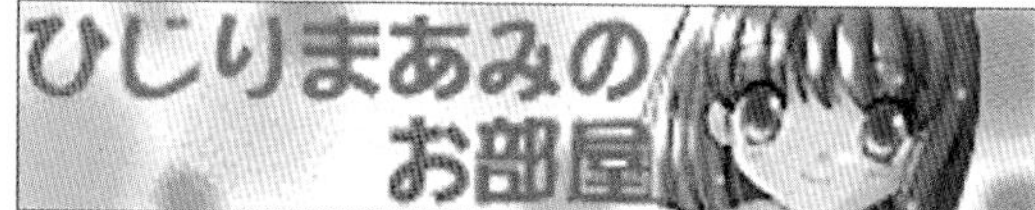

http://home2.highway.ne.jp/maami/
ひじりまあみのお部屋 / ひじりまあみ

http://home2.highway.ne.jp/maami/
ひじりまあみのお部屋 / ひじりまあみ

http://home2.highway.ne.jp/maami/
ひじりまあみのお部屋 / ひじりまあみ

http://w3ma.kcom.ne.jp/~hi6/
るるーず推進委員会 / デンタゴンるる

http://w3ma.kcom.ne.jp/~hi6/
るるーず推進委員会 / デンタゴンるる

http://ww1.tiki.ne.jp/~su-sky/
Lost Memory / す～

http://plaza18.mbn.or.jp/~s_taka/
たかはし倶楽部 / たかはし

http://plaza18.mbn.or.jp/~s_taka/
たかはし倶楽部 / たかはし

http://plaza18.mbn.or.jp/~s_taka/
たかはし倶楽部 / たかはし

http://plaza18.mbn.or.jp/~s_taka/
たかはし倶楽部 / たかはし

http://member.nifty.ne.jp/yyoshida/
魂の兄弟たち / よしいさん

yugori page
http://www.asahi-net.or.jp/~MN2Y-TBT/
yugori

登載著許多以RPG角色為題材的作品，真實地呈現出飛龍的真實感。

This site contains many images of characters with a role-playing game theme. Please have a look at this especially realistic dragon image when you visit.

http://www.kt.rim.or.jp/~eyeball/
あい・ぼうるアニメCGホームページ / あい・ぼうる

http://village.infoweb.ne.jp/~fan/
○○FANのほ~むぺ~じ! / アニック

http://village.infoweb.ne.jp/~fan/
○○FANのほ~むぺ~じ! / アニック

http://village.infoweb.ne.jp/~fan/
○○FANのほ~むぺ~じ! / アニック

http://village.infoweb.ne.jp/~katoh/
おえかきほーむぺーじ / かとうよしとも

http://kenoh.hits.ad.jp/~kenichi/
しのだよしたかのホームページ / しのだよしたか

http://www.win.ne.jp/~gonsuke/
PANDORABOX/ ごんすけ

http://village.infoweb.ne.jp/~ruby/
Ruby's Page / Ruby

http://www.bekkoame.ne.jp/~ruury/
あとりえRUURY / RUURY

http://www.bekkoame.ne.jp/~ruury/
あとりえRUURY / RUURY

http://www.bekkoame.ne.jp/~ruury/
あとりえRUURY / RUURY

http://www.bekkoame.ne.jp/~ruury/
あとりえRUURY / RUURY

http://member.nifty.ne.jp/sotokan/
CornPine / SOTOKAN

http://member.nifty.ne.jp/sotokan/
CornPine / SOTOKAN

http://plaza17.mbn.or.jp/~STrash/
Studio Trash Official Homepage / Studio Trash

http://plaza17.mbn.or.jp/~STrash/
Studio Trash Official Homepage / Studio Trash

http://ww1.tiki.ne.jp/~su-sky/
Lost Memory /す~

http://ww1.tiki.ne.jp/~su-sky/
Lost Memory /す~

http://www.ceres.dti.ne.jp/~syo/
自然を大切に / syo

http://member.nifty.ne.jp/TOKUMEIKA/
STUDIOぶーぴーとらっぷ / STUDIOぶーぴーとらっぷ

あい・ぼうるアニメCGホームページ
http://www.kt.rim.or.jp/~eyeball/
あい・ぼうる

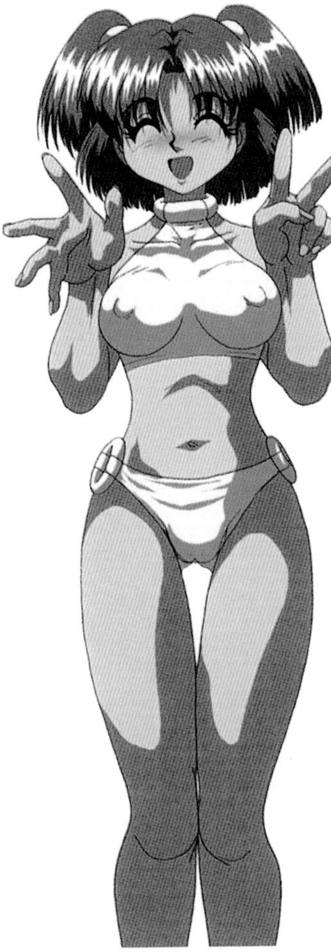

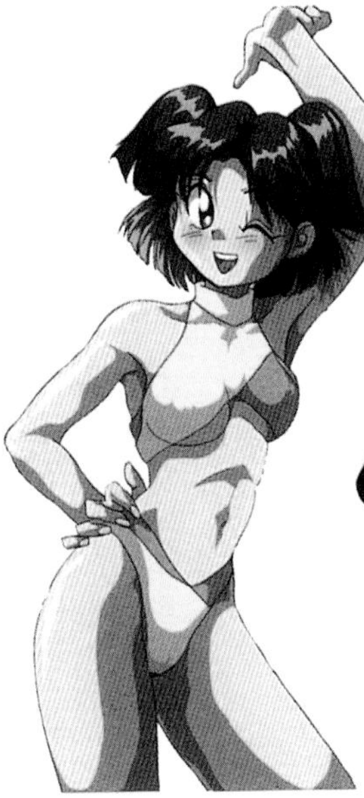

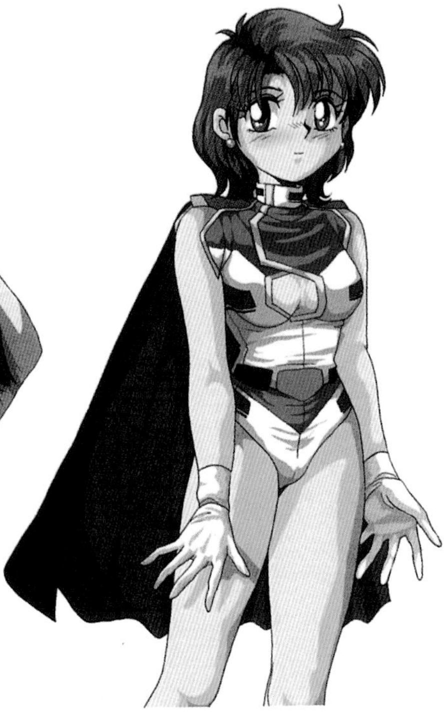

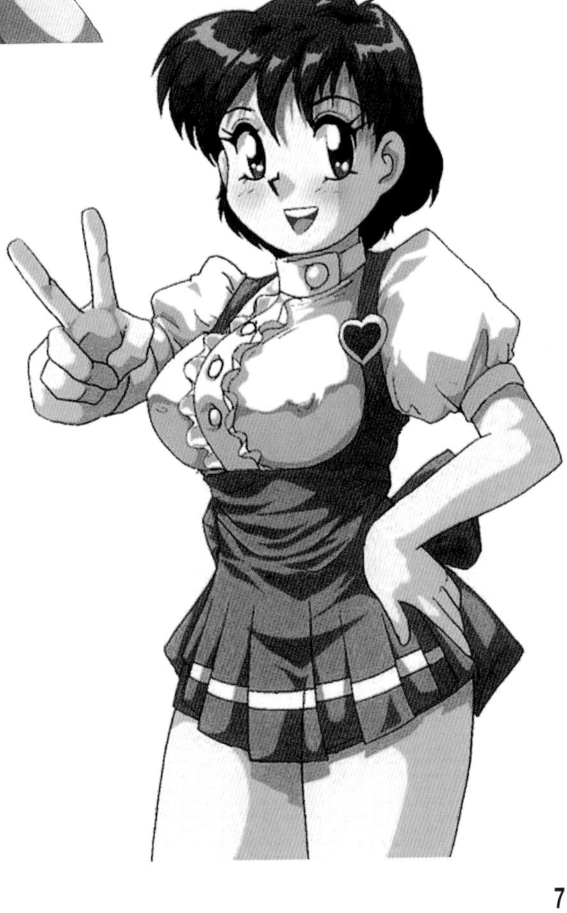

擁有200萬筆資料的CG網站，不只有最新的動畫作品，還能夠觀賞到令人懷念的動畫美少女的CG，特殊的姿態和構圖以及生氣勃勃的動態表現，這就是AI.BOULU的個人風格。

One of the most popular CG sites on the web, this site's access counter has hit over 2 million. On this site, you can enjoy CG portraits of popular TV anime characters from both popular current series and old favorites. Featured are "bishoujo" or "beautiful young girl" characters. Eyeball's specialty is cel-painting. His technique is to depict shadows using only 2 to 3 colors rather than gradiations for an animation-style look. His characters are drawn in active and suggestive poses. These three elements go together to make very pleasing compositions.

http://www.yo.rim.or.jp/~takeori/21th/
21世紀計画社・電脳社内報 / 21世紀計画社

http://www.tk.xaxon.ne.jp/~endless/
-R C A- / Endless

http://www.tk.xaxon.ne.jp/~endless/
-R C A- / Endless

http://www.tk.xaxon.ne.jp/~endless/
-R C A- / Endless

http://www.asahi-net.or.jp/~uw6n-tki/
Nobu's BambooHouse / nobu

http://jimcok.co.jp/lana/KITUPON/index.htm
KITUPON'S HP / KITUPON

http://village.infoweb.ne.jp/~imi/
ASYLUM (アサイラム) / IMI

http://plaza11.mbn.or.jp/~iris/
IRIS営業所 / IRIS

http://village.infoweb.ne.jp/~jes/
STUDIO JES HOME PAGE / JES

http://home.interlink.or.jp/~tkumagai/jinpino.html
STUDIO JIP! / JINPINO

http://www.asahi-net.or.jp/~pg4s-nw/
42番街のペテン師たち / Lips

http://miya.or.jp/~moonriya/
MOONRIYA / MOONRIYA

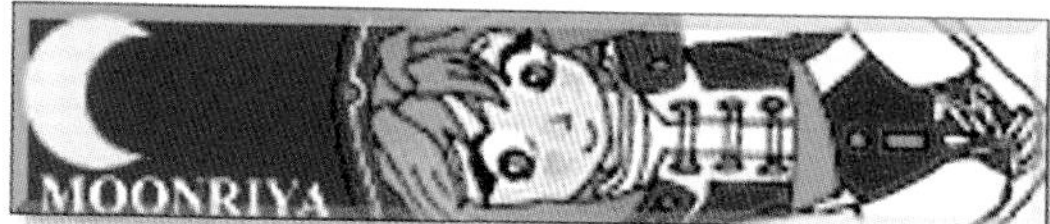

http://miya.or.jp/~moonriya/
MOONRIYA / MOONRIYA

http://m-y.ed-sys.co.jp/
Marine Web / minoru

http://www.ask.or.jp/~mao/
POT☆CRAFT / ぽてきち

http://www.yo.rim.or.jp/~takeori/sk/
SchwarzeKirschen / 21世紀計画社

http://www.yo.rim.or.jp/~takeori/sk/
SchwarzeKirschen / 21世紀計画社

http://www.yo.rim.or.jp/~takeori/sk/
SchwarzeKirschen / 21世紀計画社

http://stein.qse.tohoku.ac.jp/~hekiru/
STUDIO M / 久川 晶

http://jun.gaia.h.kyoto-u.ac.jp/~eiji/
m/art-lab / Kyo.Komiya

古宮京

m-labo

m/art-lab

http://jun.gaia.h.kyotou.ac.jp/~eiji/

Kyo.Komiya

進入頂級網頁中，不知何時已感受到小宮的魅力，那是屬於感覺的網站，其是兼具了動畫及遊戲特色，非常藝術的作品。

Starting from the opening page, this site really draws you in to keep exploring further and further. Before you know it you are completely captured by the world of the artist. Very artistic images of comic and game characters are the focus here.

http://www.geocities.co.jp/Playtown-Denei/2156/
喫茶赤べこ屋 / kami

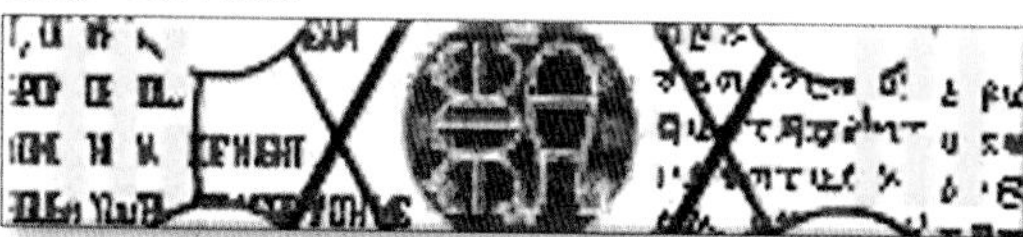
http://www.angel.ne.jp/~wawon/
朝 / 高木朝成

http://www.bekkoame.ne.jp/i/gd5293/
B-PART / たちばなふゆか

http://www.bekkoame.ne.jp/i/gd5293/
B-PART / たちばなふゆか

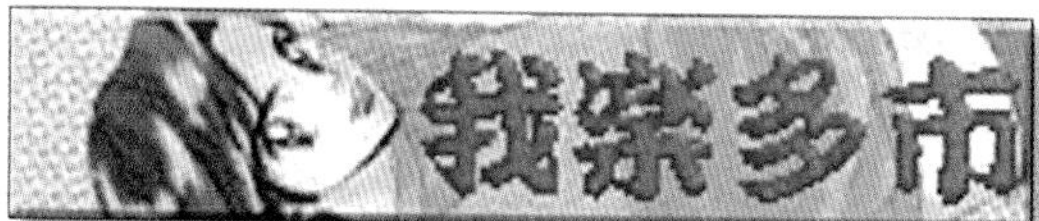

http://www.din.or.jp/~aioi-aoi/
我楽多市 / 相生葵

http://www.din.or.jp/~aioi-aoi/
我楽多市 / 相生葵

http://www.amy.hi-ho.ne.jp/taro-chiaki/
ほかほか書店 /ほかほか書店

http://web.kyoto-inet.or.jp/people/htaka/
穂高んちの作品掲示部屋 / 穂高ひとみ

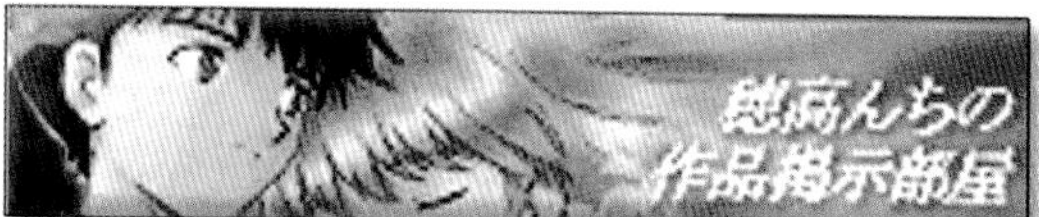

http://web.kyoto-inet.or.jp/people/htaka/
穂高んちの作品掲示部屋 / 穂高ひとみ

http://web.kyoto-inet.or.jp/people/htaka/
穂高んちの作品掲示部屋 / 穂高ひとみ

http://web.kyoto-inet.or.jp/people/htaka/
穂高んちの作品掲示部屋 / 穂高ひとみ

http://web.kyoto-inet.or.jp/people/htaka/
穂高んちの作品掲示部屋 / 穂高ひとみ

http://web.kyoto-inet.or.jp/people/htaka/
穂高んちの作品掲示部屋 / 穂高ひとみ

http://www.awave.or.jp/home/naga1615/himuraya/
緋村屋 / kami

http://nori.im.kindai.ac.jp/~myzz/
Grass Valley / 瑞杜秀智

http://nori.im.kindai.ac.jp/~myzz/
Grass Valley / 瑞杜秀智

http://nori.im.kindai.ac.jp/~myzz/
Grass Valley / 瑞杜秀智

http://nori.im.kindai.ac.jp/~myzz/
Grass Valley / 瑞杜秀智

http://www.ddt.or.jp/~shino/
しのさん美術館 / しの

http://www.ddt.or.jp/~shino/
しのさん美術館 / しの

緋村屋

http://www.awave.or.jp/home/naga1615/himuraya/

kami

登載著相當動人的作品，大膽的構圖及果斷的畫法，這是擁有獨特魅力的KAMI的網站，展現出超級動人的作品，這是我個人相當中意的網站。

Extremely dynamic images with bold colors and compositions are Kami's specialty. I really enjoy these images which are so dynamic it seems they can hardly be contained within the dimesions of the computer screen.

http://www.yk.rim.or.jp/~axis88/
AXIS WORLD / KOU

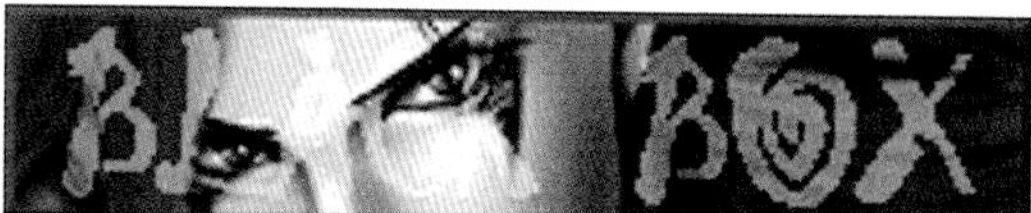

http://www.netlaputa.ne.jp/~louie/
BLACK BOX / 柴崎泪

http://www.ceres.dti.ne.jp/~yharada/
DRAGON PALACE / 原田竜介

http://www.os.rim.or.jp/~etsumi-f/
GP工房 / 藤衣 悦巳

http://www.os.rim.or.jp/~etsumi-f/
GP工房 / 藤衣 悦巳

ttp://www.try-net.or.jp/~yufu/
FLAT PACKAGE / 大東亜ゆう

ttp://www.try-net.or.jp/~yufu/
FLAT PACKAGE / 大東亜ゆう

http://www2.justnet.ne.jp/~genax/
GENAX / GENAX

http://www2.tky.3web.ne.jp/~kamkam/
Go-Q☆Factory / 風渡カムイ

http://www2.tky.3web.ne.jp/~kamkam/
Go-Q☆Factory / 風渡カムイ

http://www1.interq.or.jp/~d-hami/
HAMIDASHI 情熱系 / 間枝皇二

http://www1.interq.or.jp/~d-hami/
HAMIDASHI 情熱系 / 間枝皇二

http://www.cyborg.ne.jp/~iwaha/
仮想的小葉~Virtual Leaflet~ / いわは

http://www.bekkoame.ne.jp/i/yuduru/
JUNKBOX / 萌沢ゆづる

http://www.bekkoame.ne.jp/i/yuduru/
JUNKBOX / 萌沢ゆづる

http://www.bekkoame.ne.jp/i/yuduru/
JUNKBOX / 萌沢ゆづる

http://www.bekkoame.ne.jp/i/yuduru/
JUNKBOX / 萌沢ゆづる

http://www.bekkoame.ne.jp/~k_komuro/
STUDIO亜人類 / 小室恵佑

http://www2s.biglobe.ne.jp/~chiffon/
CHOCOLATE CHIFFON / 大阪なる

http://www.tky.3web.ne.jp/~rgb/
Circle RGB / RGB

STUDIO 亜人類

STUDIO亜人類

http://www.bekkoame.ne.jp/~k_komuro/ajinrui.html

小室恵佑

以美少女遊戲為題材，和小室同名的雜誌和CG集以及其續集作品，全都登載在這個網站上。

On this web page you can see reproductions of CG images from Keisuke's "doujin" magazine and also from his CG collection book of "bishoujo" game characters.

http://www.webnik.ne.jp/~uzumi/
A tender blue breeze / あめいすめる

http://www.geocities.co.jp/Playtown-Denei/4872/
LSD ANNEX / おじゃわ

http://www2.justnet.ne.jp/~aynm/
FLEETING HAPPINESS / 綾波

http://www.ask.or.jp/~jiji/
Jiji's Home Page / じじ

http://www.ask.or.jp/~jiji/
Jiji's Home Page / じじ

http://www.ic-net.or.jp/home/chic/
CHI倶楽部 / タカハマ

http://www.netlaputa.ne.jp/~rei-t/tazki/
Crazy Circus / 楠たつき

http://www.netlaputa.ne.jp/~rei-t/tazki/
Crazy Circus / 楠たつき

http://www.netlaputa.ne.jp/~rei-t/tazki/
Crazy Circus / 楠たつき

http://www.netlaputa.ne.jp/~rei-t/tazki/
Crazy Circus / 楠たつき

http://www.ruri.nadeshico.net/
HI-TECH CARROT / 佳由樹

http://www.e-net.or.jp/user/totorin/
LOVE SPIN DRIVE / おじゃわ

http://www.e-net.or.jp/user/totorin/
LOVE SPIN DRIVE / おじゃわ

http://www.e-net.or.jp/user/totorin/
LOVE SPIN DRIVE / おじゃわ

http://www.e-net.or.jp/user/totorin/
LOVE SPIN DRIVE / おじゃわ

http://www.e-net.or.jp/user/totorin/
LOVE SPIN DRIVE / おじゃわ

http://www.sip.or.jp/~lockhart/
manna / 朔実アンジ

http://www.alles.or.jp/~melody/
まじかるメロディ / Melody-Yoshi

http://www.fsinet.or.jp/~stms/
魁!!アニメ塾 / MASA

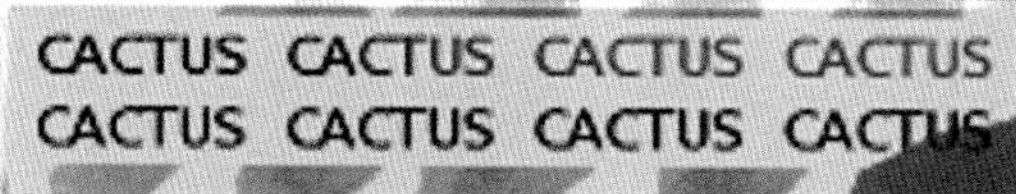

http://www.rr.iij4u.or.jp/~mirin/
CACTUS / 酒田みりん

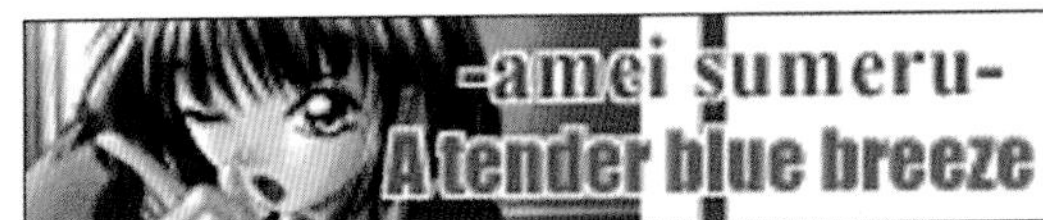

A tender blue breeze
http://www.webnik.ne.jp/~uzumi/
あめいすめる

此網站登載著將服裝、皮膚質感以及重量感表現得淋漓盡致的完美作品，果然要決定用色的確是件不容易的事，這裡登載的角色的服裝很有特色，而且款式也很多。

On this site you can see images drawn in beautiful colors and reaslistically expressed clothing and skin textures. The images seem to have a realistic weight to them instead of just being flat. The costume designs of the characters are also very cool.

http://www2.justnet.ne.jp/~rskj/
碧の館 / 海女原磨莉

http://www2.justnet.ne.jp/~rskj/
碧の館 / 海女原磨莉

http://www.asahi-net.or.jp/~SY8K-HTNK/
still I believin / RYUNE

http://www.asahi-net.or.jp/~SY8K-HTNK/
still I believin / RYUNE

http://www.asahi-net.or.jp/~SY8K-HTNK/
still I believin / RYUNE

http://www.asahi-net.or.jp/~SY8K-HTNK/
still I believin / RYUNE

http://www.asahi-net.or.jp/~SY8K-HTNK/
still I believin / RYUNE

http://www.asahi-net.or.jp/~SY8K-HTNK/
still I believin / RYUNE

http://www.asahi-net.or.jp/~SY8K-HTNK/
still I believin / RYUNE

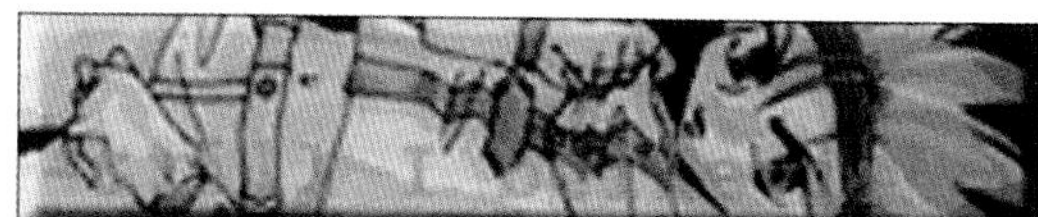
http://www.asahi-net.or.jp/~SY8K-HTNK/
still I believin / RYUNE

http://www.asahi-net.or.jp/~SY8K-HTNK/
still I believin / RYUNE

http://www.bekkoame.ne.jp/ha/konpeito/
The Critical Point Of Desire / 金米糖

http://www.gld.mmtr.or.jp/~kokai/
K.P.G / Kokai

http://www.gld.mmtr.or.jp/~kokai/
K.P.G / Kokai

http://www.gld.mmtr.or.jp/~kokai/
K.P.G / Kokai

http://www.gld.mmtr.or.jp/~kokai/
K.P.G / Kokai

http://www.gld.mmtr.or.jp/~kokai/
K.P.G / Kokai

http://www.gld.mmtr.or.jp/~kokai/
K.P.G / Kokai

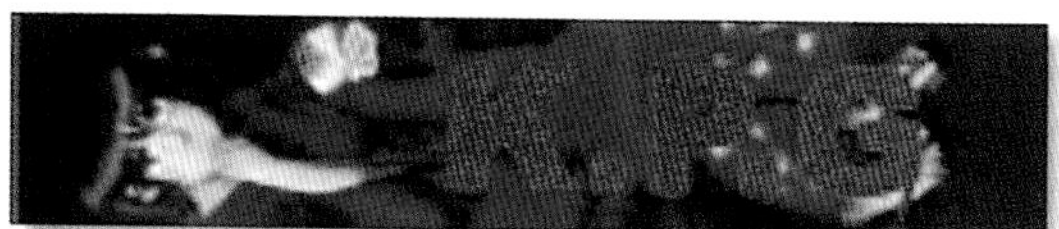
http://www.gld.mmtr.or.jp/~kokai/
K.P.G / Kokai

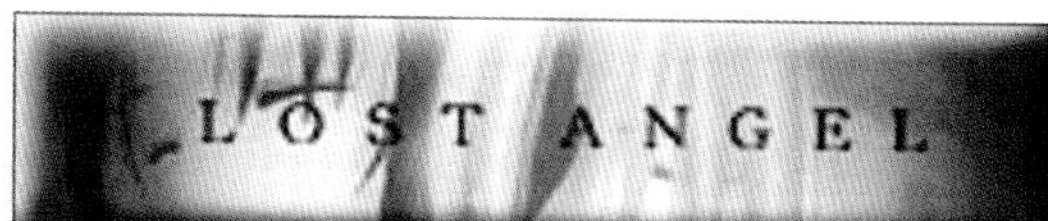

http://www1.u-netsurf.or.jp/~mizuki/lost/
Lost Angel / みずき

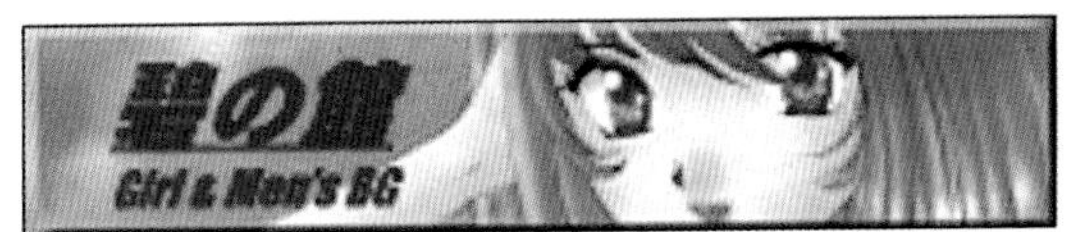

碧の館
http://www2.justnet.ne.jp/~rskj/
海女原磨莉

上述的URL是通往海女原三個網站的重要通道，所有以遊戲及動畫為題材的作品，便以網站來區分，共可區分出格鬥遊戲、櫻花大戰等等，更有趣的是CG部份也區分出「美少女館」及「美男子館」。

The above URL leads you to an entrance page for three sites. All three sites feature images of game and anime characters. The different sites are, "Fighting Soul" which has characters from fighting games, "Eeizen Kreug", a Sakura Wars page and "Aonoyakata" which has both "bishoujo" and "bidanshi" (beautiful young man) images.

http://www7.peanet.ne.jp/~cho/
Pastime / ちょも

http://www7.peanet.ne.jp/~cho/
Pastime / ちょも

http://www.cc.rim.or.jp/~ocha/
Drawing Factor / OCHA

http://www.tk.airnet.ne.jp/kotone_h/
Henachocosystem On-Line / Project katz

http://www3.osk.3web.ne.jp/~yonehara/
JG-7 / Ryu-Akt

http://www.mars.dti.ne.jp/~ka2da/
Ka2DA Factory / Ka2DA

http://www.netlaputa.ne.jp/~kenix/
E-KENIX / にんにん!

http://www.netlaputa.ne.jp/~kenix/
E-KENIX / にんにん!

http://www.sun-inet.or.jp/~urakan/
L·KAN研究所 / うらかんRN

http://www.valley.ne.jp/~occult/
Oh! CULT Room / おかると

http://www02.so-net.ne.jp/~mkmk/
MK2 FACTORY / めけめけ

http://www.kk.iij4u.or.jp/~moto/
MOTO'S CG ROOM / MOTO

http://www02.so-net.ne.jp/~mumu/
MUMU屋本舗 / MUMU

http://www.freepage.total.co.jp/hatuneandmulti/
MUTIPETADOKYUN!! / かなこ

http://www.din.or.jp/~miu/
Mutation / みゅう

http://www.geocities.co.jp/Playtown/8106/
NegEdge / takoyaki

http://www2.wbs.ne.jp/~system9/
Propaganda / 九石はくね

http://www2.wbs.ne.jp/~system9/
Propaganda / 九石はくね

http://www2.wbs.ne.jp/~system9/
Propaganda / 九石はくね

Drawing Factor
http://www.cc.rim.or.jp/~ocha/
OCHA

(C) 1997, M.M.K., Drawing Factor, OCHA

從原創角色作品中，到以動畫或人物特寫為主題的作品，全都收錄在這個網站中，從第一頁就可直接進入其他作品的網頁，進而組成了一個很有趣的網頁。

This site contains various different kinds of images from original characters designs to anime and super hero images. From the top page you can access every images directly.

http://www.yk.rim.or.jp/~as1130/
Untitled-1 / As'257G

http://www.mars.dti.ne.jp/~sato-p/
SPR NEWS / さとP

http://www2e.biglobe.ne.jp/~ichise/
Yagiyama Publishing / いちせ

http://www.lares.dti.ne.jp/~dfa/pit/
pit_inn / ぴぃ♪

http://www2s.biglobe.ne.jp/~plumroom/
plumroom / plum

http://www.enjoy.ne.jp/~jyk/
POWWOW / 樹華

http://www.ggf.ilc.or.jp/user0/retu/
Retu's Home Page Cocktail / 比十之 烈

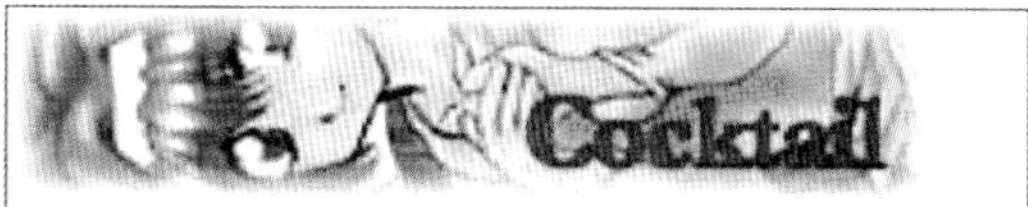

http://www.ggf.ilc.or.jp/user0/retu/
Retu's Home Page Cocktail / 比十之 烈

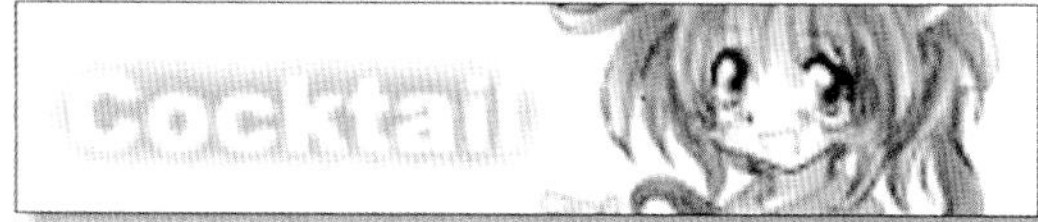

http://www.ggf.ilc.or.jp/user0/retu/
Retu's Home Page Cocktail / 比十之 烈

http://www.yk.rim.or.jp/~jaja/
The Little World of jaja / じゃじゃ

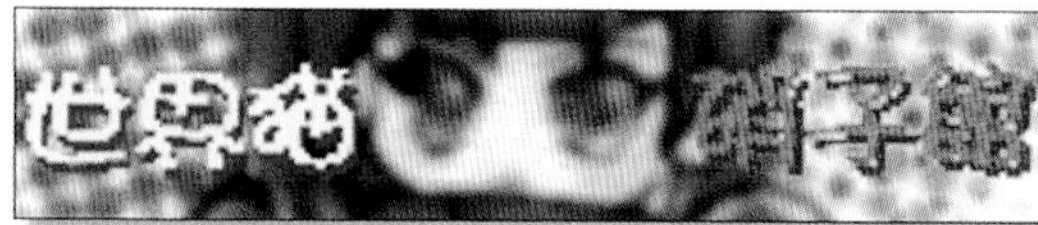

http://www.meix-net.or.jp/~noah/
世界猫硝子館 / 遊魔海里

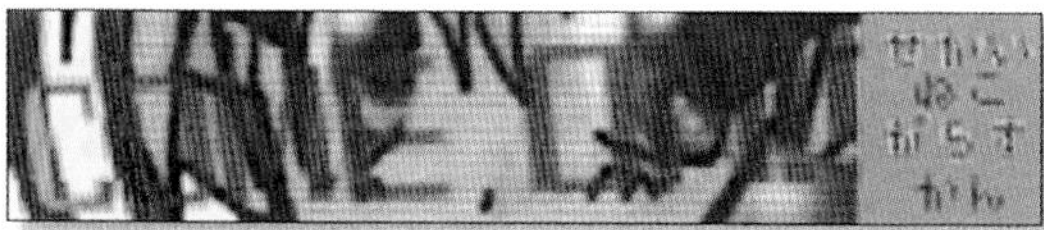

http://www.meix-net.or.jp/~noah/
世界猫硝子館 / 遊魔海里

http://www.meix-net.or.jp/~noah/
世界猫硝子館 / 遊魔海里

http://www.meix-net.or.jp/~noah/
世界猫硝子館 / 遊魔海里

http://www.meix-net.or.jp/~noah/
世界猫硝子館 / 遊魔海里

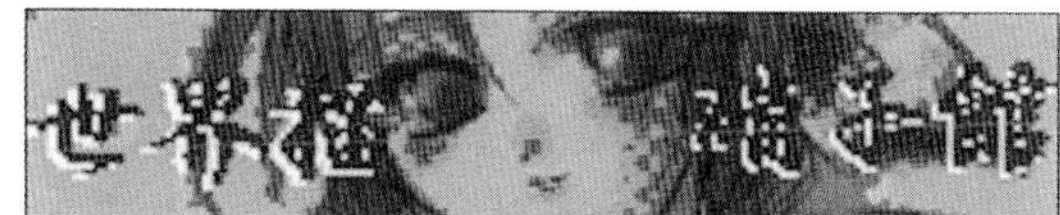

http://www.meix-net.or.jp/~noah/
世界猫硝子館 / 遊魔海里

http://www.sanynet.ne.jp/~n-n-e/
SITE-COCO / コイデココロ

http://www.netlaputa.ne.jp/~takas/
TAKA's on Web. / TAKAHiCo

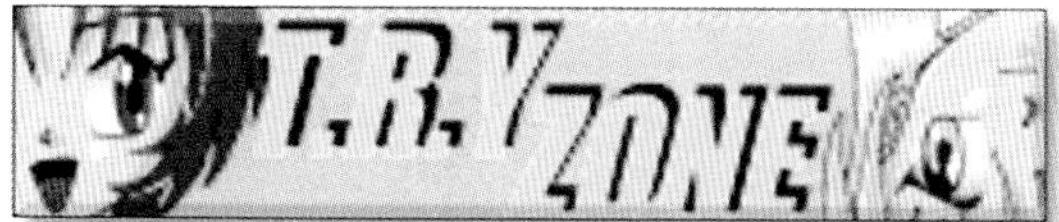

http://www.sun-inet.or.jp/~uiy00810/
T.R.Y-ZONE / 羅陰知英

http://www2.117.ne.jp/~unnet/
UN⇔NET / 雲竹彩一N

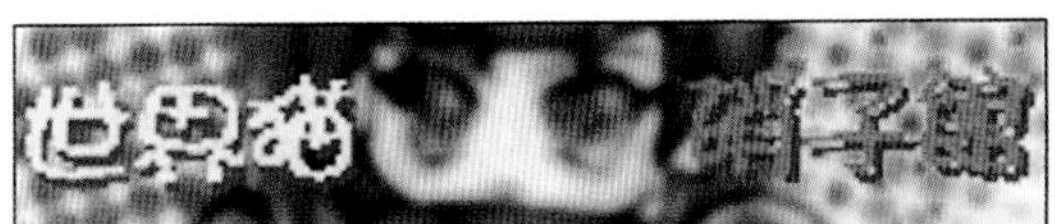

世界猫硝子館
http://www.meix-net.or.jp/~noah/
遊魔海里

這是FF main網站，作者-遊魔也公開自作的TRPG小說，其中因使用精靈而造成相當的風行，另外登載在網站中也引起了大眾相當廣泛的討論，而本書能將召喚魔法的光完美地印刷出,真是太棒了。

This is the main site for Final Fantasy (FF). The web author also presents her TRPG (role-playing game) novels here. Grace, the character shown here, is a spirit controlling magician from one of her novels. I wish the light Grace is using to summon the spirits could be printed as brightly as in the original CG image.

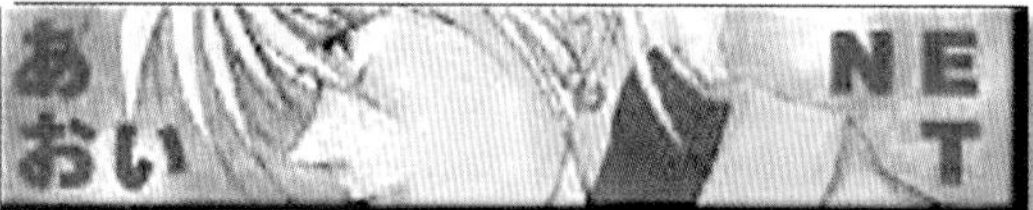

http://www3.justnet.ne.jp/~mikas/
あおいNET / MIKAMIKA

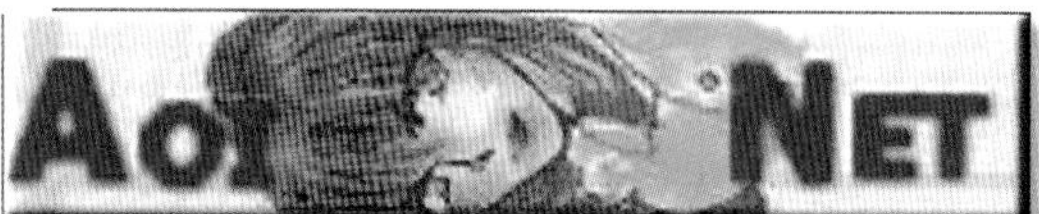

http://www3.justnet.ne.jp/~mikas/
あおいNET / MIKAMIKA

http://www.tokyo-bay.or.jp/~yamane/
BLUE BLACK / 葉月

http://www.tokyo-bay.or.jp/~yamane/
BLUE BLACK / 葉月

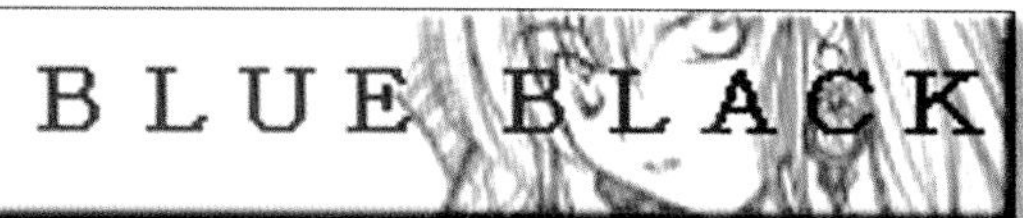

http://www.tokyo-bay.or.jp/~yamane/
BLUE BLACK / 葉月

http://www.tokyo-bay.or.jp/~yamane/
BLUE BLACK / 葉月

http://www.tokyo-bay.or.jp/~yamane/
BLUE BLACK / 葉月

http://www.tokyo-bay.or.jp/~yamane/
BLUE BLACK / 葉月

http://www.tokyo-bay.or.jp/~yamane/
BLUE BLACK / 葉月

http://www.os.xaxon.ne.jp/~spriggan/
スタジオ DELTA / SPRIGGAN

http://www.os.urban.ne.jp/home/hayasaka/
cute core / 早坂奈槻

http://www.fsinet.or.jp/~yuris_c/
Yuris cafe / 椎名悠理

http://www.fsinet.or.jp/~yuris_c/
Yuris cafe / 椎名悠理

http://www.alles.or.jp/~takew/
Site Take-W / 武W

http://www.alles.or.jp/~takew/
Site Take-W / 武W

http://www.cute.or.jp/~makuchan/
まどかぱ〜く! / 膜ちゃん

http://www.ra.sakura.ne.jp/~pelmo/madoka/
まどかぱ〜く！別館 / 膜ちゃん

http://www.geocities.co.jp/Playtown/3603/
KOCHAN STUDIO ジオシティーズ営業所 / KOCHAN

http://www.geocities.co.jp/Playtown/3603/
KOCHAN STUDIO ジオシティーズ営業所 / KOCHAN

KOCHAN STUDIO ジオシティーズ営業所
http://www.geocities.co.jp/Playtown/3603/
KOCHAN

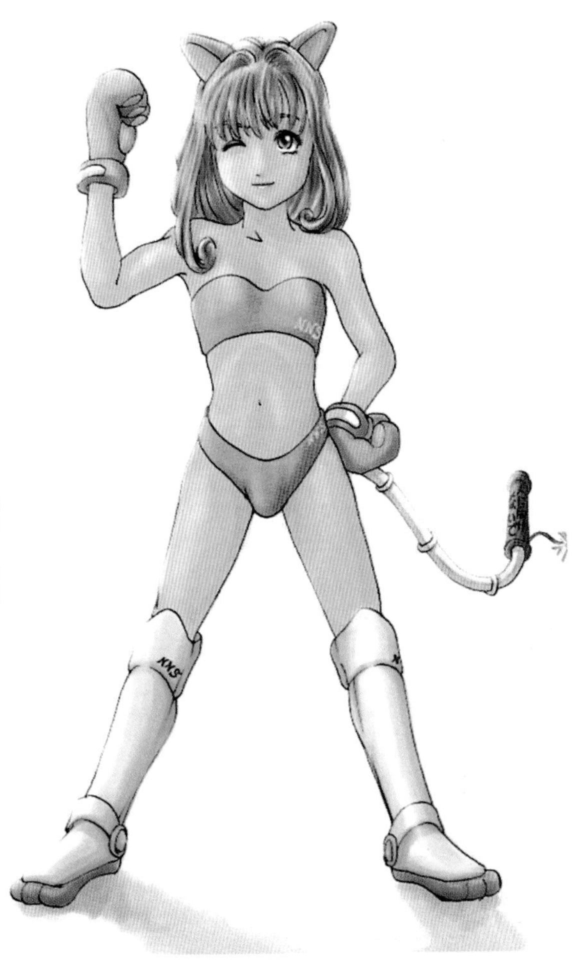

其除了公開個人風格的作品之外，還有其他以遊戲及動畫為題材的作品，在網站中還登載著原創作性的作品，因為主旨相當有趣，所以一定要造訪這個網站喔！

In addition to anime and game images, very well drawn original images are shown here. For my book, some of the original images are featured. The design of this home page is very interesting so I recommend you to visit this page.

http://www.vc-net.or.jp/~tosiyuki/
To. Digital ARTs. / To.

http://www.sa.sakura.ne.jp/~isomec/
iSOMEC's Home-Page / iSOMEC

http://www.sa.sakura.ne.jp/~isomec/
iSOMEC's Home-Page / iSOMEC

http://www.mars.dti.ne.jp/~ka2da/labor_girl/index_lg.html
レイバー少女産業振興協会 / Ka2DA

http://www.alles.or.jp/~inukai/
Metal Dress / 犬飼 紅輝

http://village.infoweb.ne.jp/~saeki/
瓢箪本舗 / 佐伯涼子

http://village.infoweb.ne.jp/~saeki/
瓢箪本舗 / 佐伯涼子

http://village.infoweb.ne.jp/~saeki/
瓢箪本舗 / 佐伯涼子

http://village.infoweb.ne.jp/~saeki/
瓢箪本舗 / 佐伯涼子

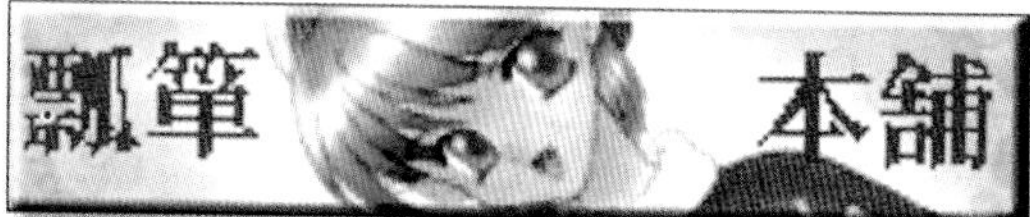

http://village.infoweb.ne.jp/~saeki/
瓢箪本舗 / 佐伯涼子

http://www.dokidoki.ne.jp/home2/ayako/
Water Colors / ayako

http://www.dokidoki.ne.jp/home2/ayako/
Water Colors / ayako

http://www.dokidoki.ne.jp/home2/ayako/
Water Colors / ayako

http://www.dokidoki.ne.jp/home2/ayako/
Water Colors / ayako

http://www2s.biglobe.ne.jp/~don-a/
Z・X・カンパニー / DON

http://www2s.biglobe.ne.jp/~don-a/
Z・X・カンパニー / DON

http://club.pep.ne.jp/~hatahira/
学園はにもくお / 学園はにもくお

iSOMEC's Home-Page
http://www.sa.sakura.ne.jp/~isomec/
iSOMEC

除了公開個人風格的作品外，還有以格鬥角色等為題材的作品，他對能繪出TNG角色一事，感到非常高興，這些都是非常珍貴的作品。

Besides original images, this site has images with scenes from the anime series "Eva", "Sailor Moon" and also fighting game characters. Personally I was very glad to see the TNG characters also here.

http://www.alles.or.jp/~aim/
G3 Aim'sHomePage / Aim

http://www.bekkoame.ne.jp/i/fl2106/
天使の涙 / くたぼでぃ

http://www.asahi-net.or.jp/~zj3y-situ/cgroom.html
滅砕CGROOM / JMS

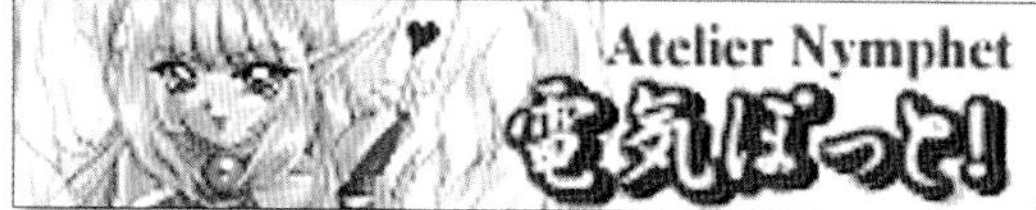

http://www.bekkoame.ne.jp/i/fl2179/
電気ポット / 攻牙沙

http://www.bekkoame.ne.jp/i/fl2179/
電気ポット / 攻牙沙

http://www.bekkoame.ne.jp/i/fl2179/
電気ポット / 攻牙沙

http://www.bekkoame.ne.jp/i/fl2179/
電気ポット / 攻牙沙

http://home3.highway.ne.jp/banana/
へっぽこギャラリー / りとるぐれい

http://home3.highway.ne.jp/~huan/
STUDIOふあん / 来鈍

http://www.bekkoame.ne.jp/~perpasia/index.html
烏賊川通信社 / perpasia

http://www.aianet.ne.jp/~yamabi/
霞草館 / 山火 霞

http://www.bekkoame.ne.jp/i/kumagon/
とった～くんの部屋 / 熊木十志和

http://www.bekkoame.ne.jp/i/kumagon/
とった～くんの部屋 / 熊木十志和

http://www.alles.or.jp/~easter/
It's a KATU'S HOMEPAGE!! / KATU

http://www.alles.or.jp/~easter/
It's a KATU'S HOMEPAGE!! / KATU

http://www.bekkoame.ne.jp/~utty0/
秘密結社Neo-Sea-Hoese / 魔導師うっちー

http://home.interlink.or.jp/~ryo1/
Ryo's Collection / Ryo

http://home3.highway.ne.jp/pachi-2/
PACHI PACHI / 天野かおる

PACHI PACHI

http://home3.highway.ne.jp/pachi-2/

天野かおる

天野是非常活躍的職業漫畫家。在網站上也可欣賞到其以動畫和遊戲角色為題材的作品。

Kaoru is a professional manga creator. This site displays Kaoru's images of anime and game characters.

http://www.bekkoame.ne.jp/~a_eu/
ありすの喫茶店AOU(E)UNIT / 緒空賀葵

http://www.bekkoame.ne.jp/~d_works/
mu genera / 士門

http://www.aitech.ac.jp/wing/~meaken/gyouchu/welcome/
暁宙館★☆ / 暁宙

http://www.top.or.jp/~kiran/kenkoku/
建国計画 / 騎羅

http://member.nifty.ne.jp/YUME/
此処あ夢工房 /夢工房

http://members.xoom.com/okada/
KAZUBOH Art Garden / KAZUBOH

http://members.xoom.com/okada/
KAZUBOH Art Garden / KAZUBOH

http://members.xoom.com/okada/
KAZUBOH Art Garden / KAZUBOH

http://members.xoom.com/okada/
KAZUBOH Art Garden / KAZUBOH

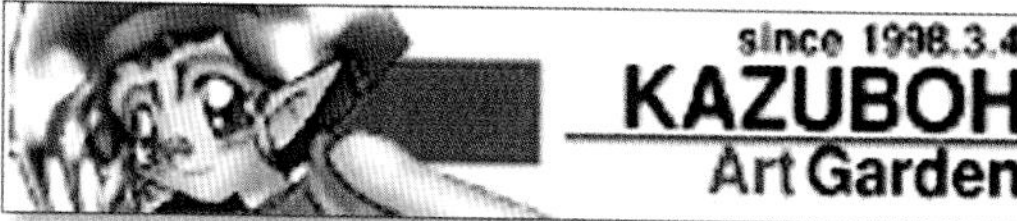

http://members.xoom.com/okada/
KAZUBOH Art Garden / KAZUBOH

http://plaza8.mbn.or.jp/~gosui/
しめじのまぜごはん / 秋野しめじ

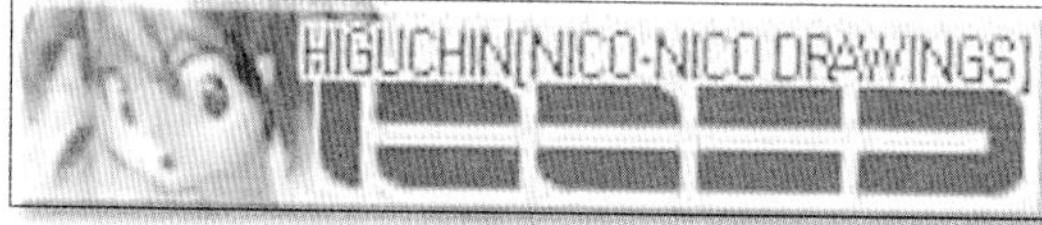

http://home3.highway.ne.jp/higuchin/
HIGUCHIN [にこニコ DRAWINGS] / HIGUCHIN

http://web1.tinet-i.ne.jp/user/chinke/
PING's WING / ぴんぐ

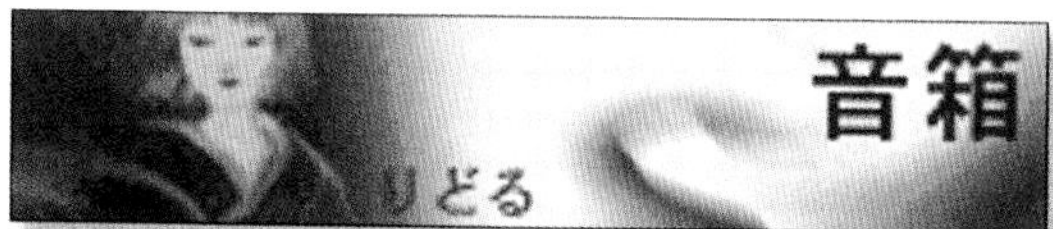

http://www.bekkoame.ne.jp/ro/ridoru/
音箱 / りどる

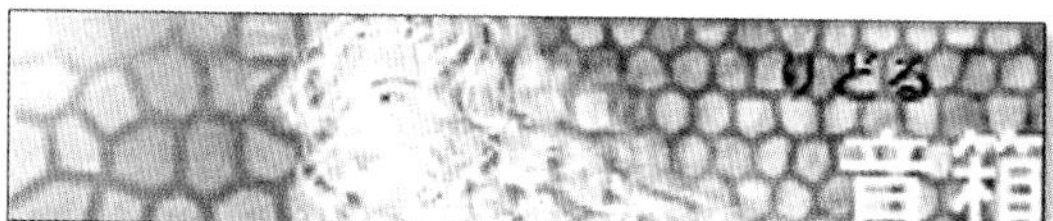

http://www.bekkoame.ne.jp/ro/ridoru/
音箱 / りどる

http://member.nifty.ne.jp/Yu-Ki/
SnowHouse / ユキ

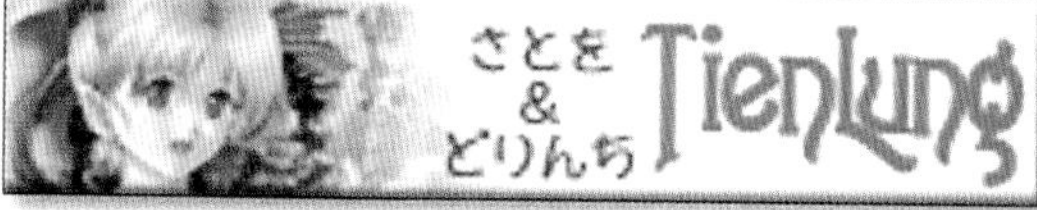

http://plaza29.mbn.or.jp/~tienlung/
T'ienLung (ティエンルン) / さとをみどり、どりんち

http://plaza29.mbn.or.jp/~tienlung/
T'ienLung (ティエンルン) / さとをみどり、どりんち

http://member.nifty.ne.jp/usagiya/
うさぎ屋本舗 / 妹尾ゆふ子

http://www.asahi-net.or.jp/~ub3k-twr/
STUDIO VIEWPORT /水瀬晶史

PING'S WING

PACHI PACHI
http://home3.highway.ne.jp/pachi-2/
天野かおる

有許多宛如天使般美少女的作品，從眉目之間流露出不同的情感，您一定會喜歡，PINK也是一位對色彩有其獨到見解的人。

This site has many images of winged girl characters. I like these images because I can clearly tell the characters' emotions from the expressions of their eyes and the way they purse their lips. Ping is also another artist who uses very original color choices.

http://www.mars.dti.ne.jp/~aki-m/
INTERNET STUDIO AKI / 三沢秋太郎

http://user.shikoku.ne.jp/hmx12mlt/
風の森工房 / マサ

http://user.shikoku.ne.jp/hmx12mlt/
風の森工房 / マサ

http://www.fang.or.jp/~bahamut/
Bahamut's Garden / Bahamut

http://www.fang.or.jp/~bahamut/
Bahamut's Garden / Bahamut

http://www.fang.or.jp/~bahamut/
Bahamut's Garden / Bahamut

http://www.fang.or.jp/~bahamut/
Bahamut's Garden / Bahamut

http://www.fang.or.jp/~bahamut/
Bahamut's Garden / Bahamut

http://www.fang.or.jp/~bahamut/
Bahamut's Garden / Bahamut

http://www2.justnet.ne.jp/~h_yuse/
YUSER'S倶楽部 / TM・ゆ〜ず

http://www2.justnet.ne.jp/~h_yuse/
YUSER'S倶楽部 / TM・ゆ〜ず

http://www2.justnet.ne.jp/~h_yuse/
YUSER'S倶楽部 / TM・ゆ〜ず

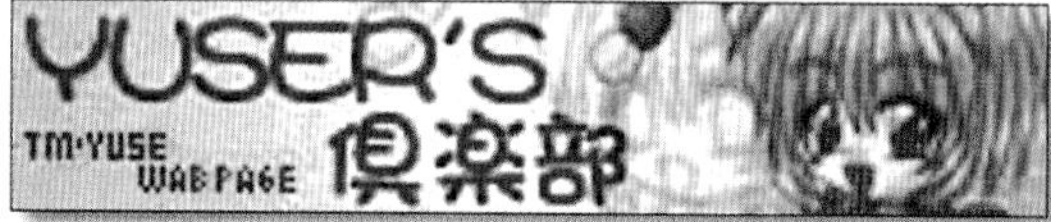

http://www2.justnet.ne.jp/~h_yuse/
YUSER'S倶楽部 / TM・ゆ〜ず

http://www.ops.dti.ne.jp/~ive/
Obsidian Wind JP / ive

http://www.ops.dti.ne.jp/~ive/
Obsidian Wind JP / ive

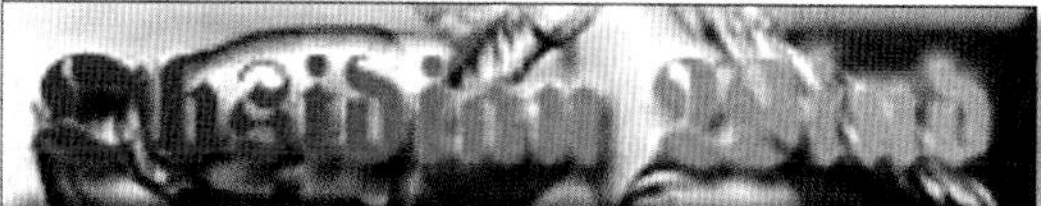

http://www.ops.dti.ne.jp/~ive/
Obsidian Wind JP / ive

http://www.ops.dti.ne.jp/~ive/
Obsidian Wind JP / ive

http://mypage.goplay.com/obs/
Obsidian Wind USA / ive

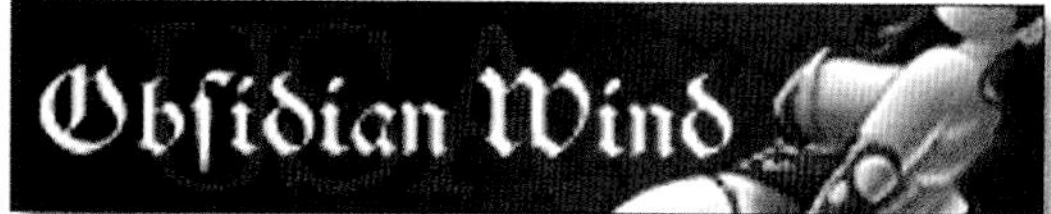

http://mypage.goplay.com/obs/
Obsidian Wind USA / ive

Bahamut's Garden
http://www.fang.or.jp/~bahamut/
Bahamut

Bahamut網站登場的是以幻想世界的小妖精和亞人種為題材的作品。附有IE4.0限定指南，各位一定要一探究竟。

From Bahamut's web site you can enjoy many images of elf-type fantasy world characters. The interface index works best with Internet Explorer 4.0. This index uses BGM for aural guidance which I think is very cool and a must see.

http://www2.marinet.or.jp/~sakuraba/
DayDream / 碧竜ヲサム

http://www2.marinet.or.jp/~sakuraba/
DayDream / 碧竜ヲサム

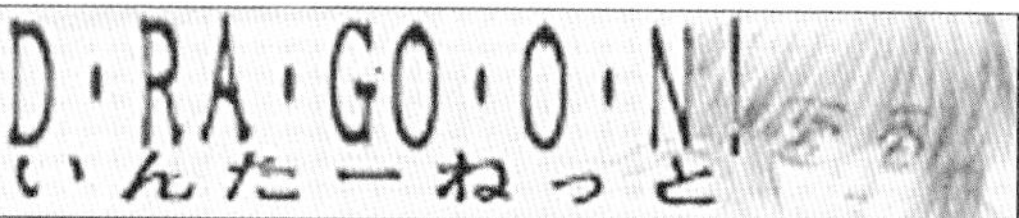

http://www.kt.rim.or.jp/~youie/
D・RA・GO・O・N! いんたーねっと / 結維

http://www.kt.rim.or.jp/~youie/
D・RA・GO・O・N! いんたーねっと / 結維

http://www.kumagaya.or.jp/~koju/
Esel神聖帝国 / Esel

http://www.kumagaya.or.jp/~koju/
Esel神聖帝国 / Esel

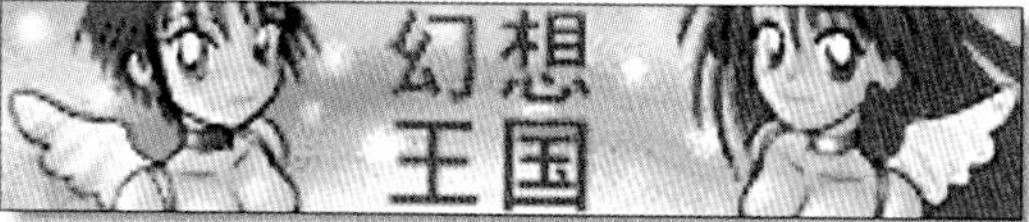

http://www.alles.or.jp/~funny/
幻想王国 / Funny.

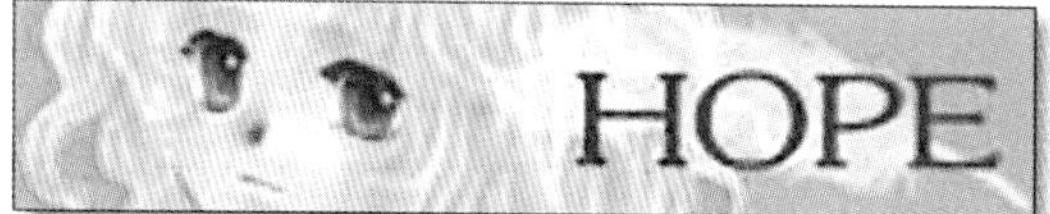

http://www.campus.ne.jp/~takano/
HOPE / はるの

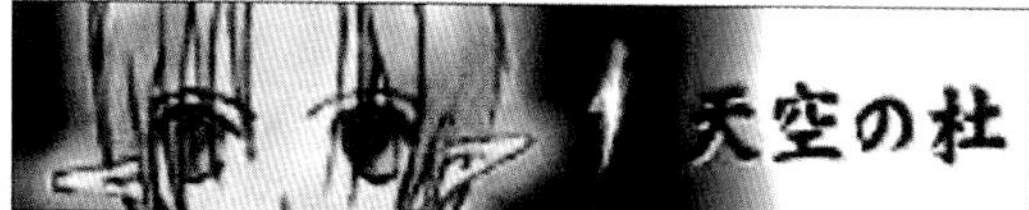

http://club.pep.ne.jp/~apt.hosono/index.html
天空の杜 / 智樹

http://club.pep.ne.jp/~apt.hosono/index.html
天空の杜 / 智樹

http://www.bekkoame.ne.jp/~guernica/
海カラス共和国 / ゲルニカ

http://www3.big.or.jp/~bon/
Gallery ILLUSION / BON

http://www3.big.or.jp/~bon/
Gallery ILLUSION / BON

http://www.sp.dianet.or.jp/~pai/
HomePage PAI / PAI

http://www02.u-page.so-net.ne.jp/xb3/ryu-sei/
R Project / 瑠堂れおな

http://www.din.or.jp/~shura/
TOYPOP IN WAVE / shura

http://member.nifty.ne.jp/riy/
惑星たごや / 竜にょ

http://member.nifty.ne.jp/riy/
惑星たごや / 竜にょ

http://user.shikoku.ne.jp/hmx12mlt/
風の森工房 / マサ

http://member.nifty.ne.jp/yositaka/
星影美術館 / 濱口よしたか

Gallery ILLUSION
http://www3.big.or.jp/~bon/
BON

BON網站中登載的，都是充滿幻想以及陽剛性很強的插畫，在非常雅緻的插畫中，出現了陽剛氣十足的作品時，真是另人宛然一笑。

This site displays original fantasy illustrations with many macho "aniki" older brother type characters. When I found the "Machona Taiyou" (macho sun) image among other images with a very quiet and peaceful taste, I burst out laughing.

http://www.win.ne.jp/~tomochan/3g.html
3G / Joe

http://www.synapse.ne.jp/~bancho/
-番長列伝- MIG-29's Home Page / MIG-29

http://www2.odn.ne.jp/~caj16480/
icenou / icenou

http://www.linkclub.or.jp/~mickey/
MICKEY'S LAGOON / MICKEY

http://www.zokei.asu.ac.jp/~nbu/
NBU 3DCG GALLERY / HIROPU

http://www.zokei.asu.ac.jp/~nbu/
NBU 3DCG GALLERY / HIROPU

http://www.zokei.asu.ac.jp/~nbu/
NBU 3DCG GALLERY / HIROPU

http://www.asahi-net.or.jp/~fv3n-wkby/
3DCG補完委員会 / WAKA

http://www02.so-net.ne.jp/~vincho/
緒計画 / VIN緒

http://www.asahi-net.or.jp/~fv3n-wkby/
3DCG補完委員会 / WAKA

http://www.aikis.or.jp/~s-iwao/
がんちゃんの3DCGのページ / 岩尾眞吾

http://www.alles.or.jp/~honey/
ハニー ソノちゃま らぶらぶっこ倶楽部 / 三上 ソノ

http://www02.so-net.ne.jp/~yukine/
STUDIO SUNNY SPOT / ひなた・ゆきね

http://www.asahi-net.or.jp/~je6t-fjt/tap/
TAP WORLD / たっぷ

http://www.asahi-net.or.jp/~je6t-fjt/tap/
TAP WORLD / たっぷ

http://www.asahi-net.or.jp/~SD3T-KTU/
TK150's Miscellaneous Area / TK150

http://www3.osk.3web.ne.jp/~tosino/
ToSINo Gallery / ToSINo

http://stu.nit.ac.jp/~e977006/
VANGUARD FLIGHT / 押野 卓

http://stu.nit.ac.jp/~e977006/
VANGUARD FLIGHT / 押野 卓

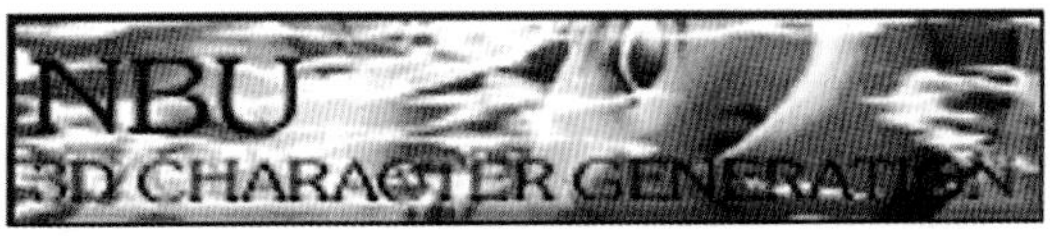
NBU
3D CHARACTER GENERATION

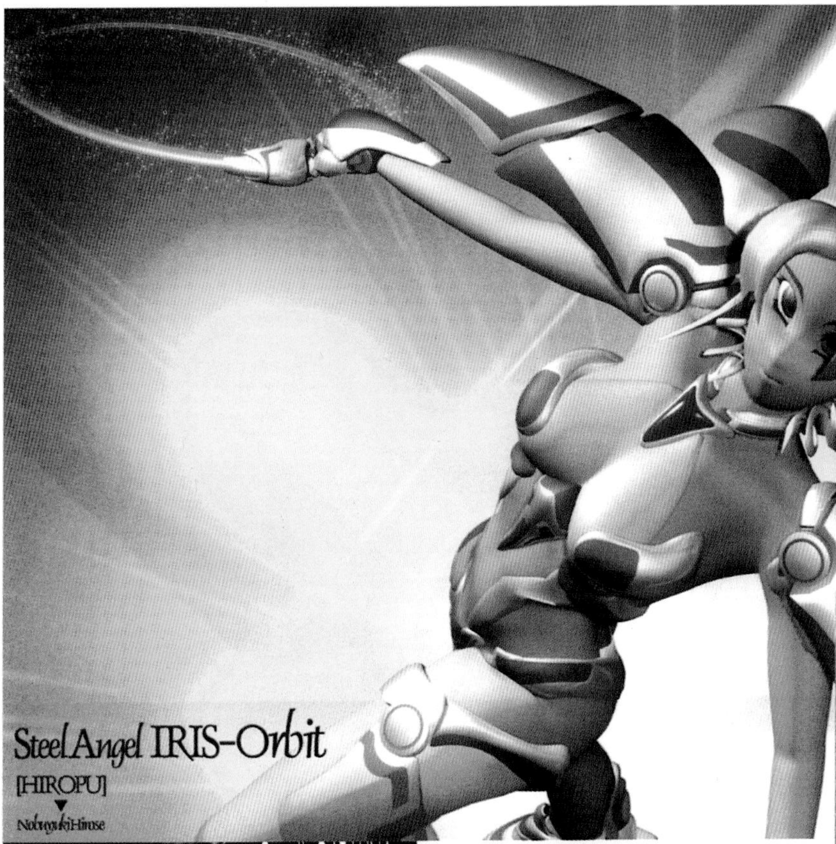
Steel Angel IRIS-Orbit
[HIROPU]
Nobuyuki Hirose

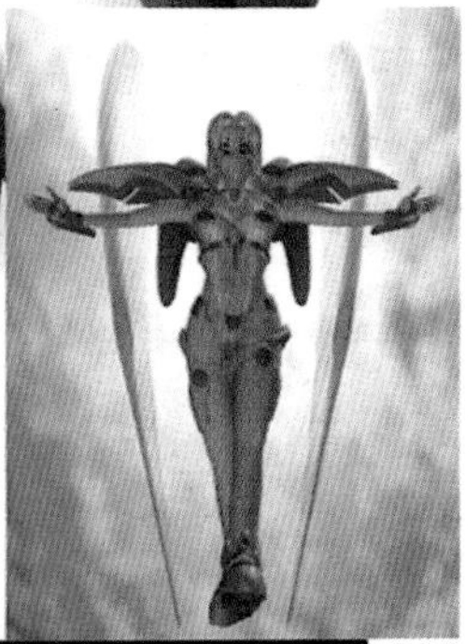

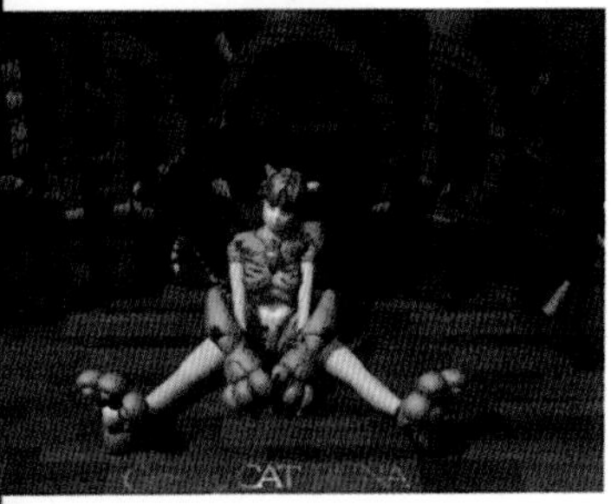
GravityCAT LUNA

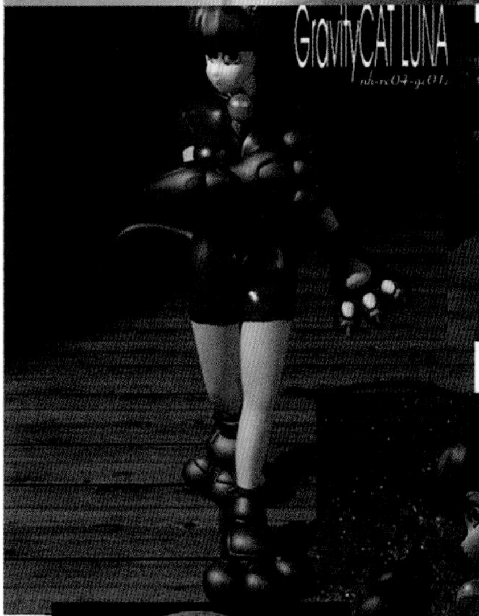

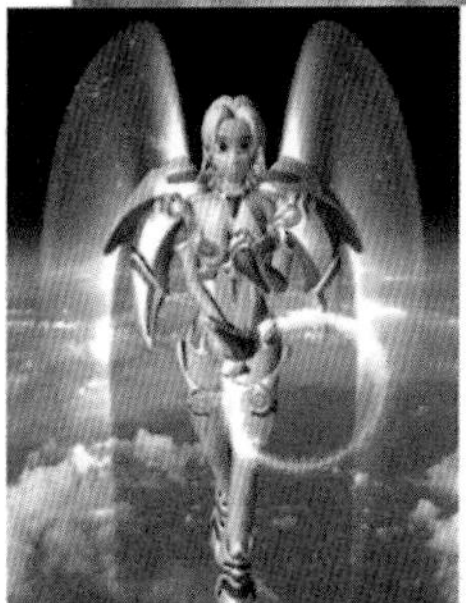

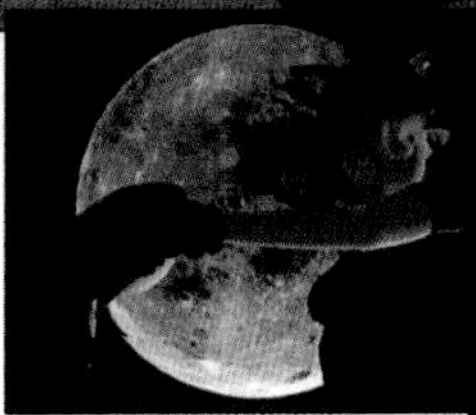

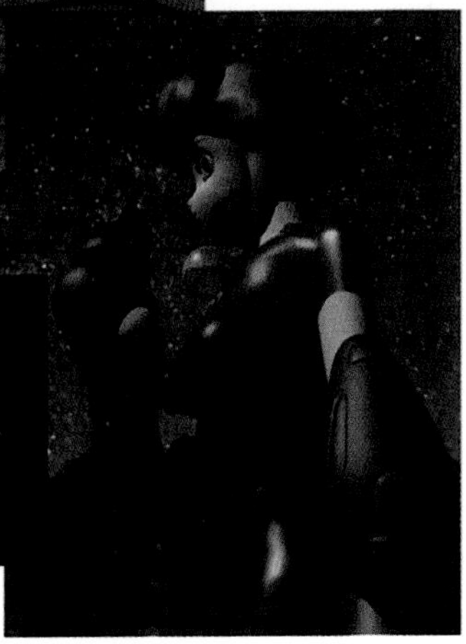

HIROPU網站登載了許多以3D繪圖工具所製作而成的作品。他提供給我描繪可愛女子角色的作品，而他也對能在電腦上將角色立體化一事，感到興奮莫名。

Hiropu's web site features images made using 3D graphics tools. Out of these, this pretty girl image was chosen to appear in the book. It's really exciting to imagine that someone's original character design idea can transformed into a 3D figure on the computer screen like this.

http://www.bekkoame.ne.jp/ro/haiboku/
めがねがね / はいぼく

http://plaza12.mbn.or.jp/~horaku/
豊楽煩悩館 / 豊楽

http://www.alpha-net.or.jp/users/sakujima/
上海楼 / 佐久嶋火急

http://www.bekkoame.ne.jp/i/kaede/
kaedeはうす / 楓 T2

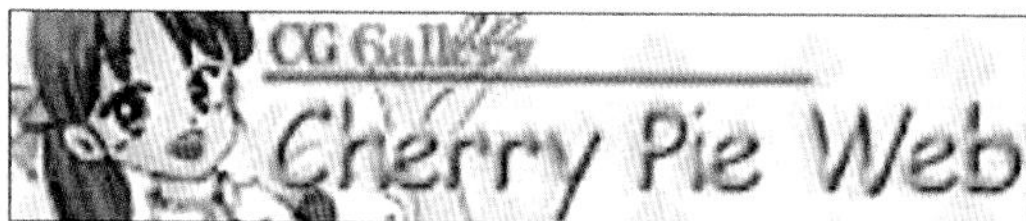

http://www.vector.co.jp/authors/VA004639/
Cherry Pie Web / さくらぎけい

http://www.geisya.or.jp/~kiyo/
○急電鉄 / きよ○

http://www.geisya.or.jp/~kiyo/
○急電鉄 / きよ○

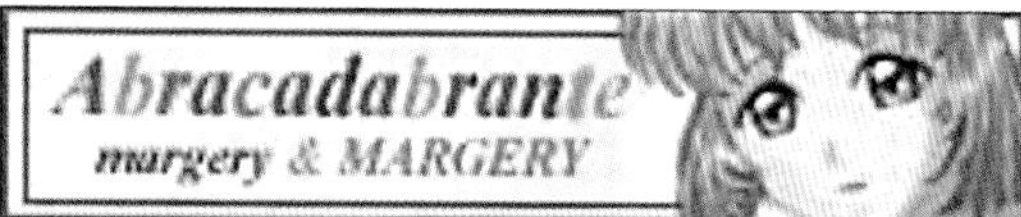

http://www.netwave.or.jp/~margery/
Abracadabrante margery & MARGERY / margery

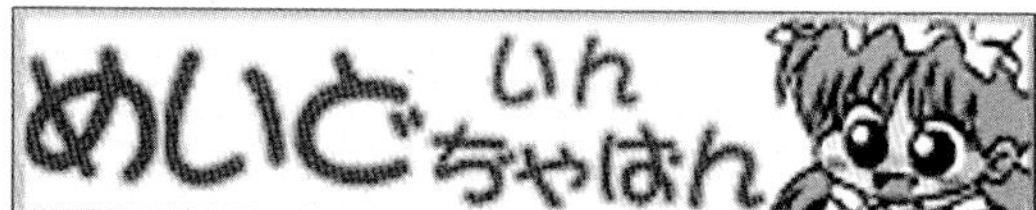

http://www.asahi-net.or.jp/~nu7h-ootw/
めいどいんちゃばん / 眉毛

http://www.alles.or.jp/~msuigun/
村上水軍の館 / 村上水軍

http://www.alles.or.jp/~msuigun/
村上水軍の館 / 村上水軍

http://member.nifty.ne.jp/marui/
Forza! / のぶまる

http://member.nifty.ne.jp/marui/
Forza! / のぶまる

http://member.nifty.ne.jp/marui/
Forza! / のぶまる

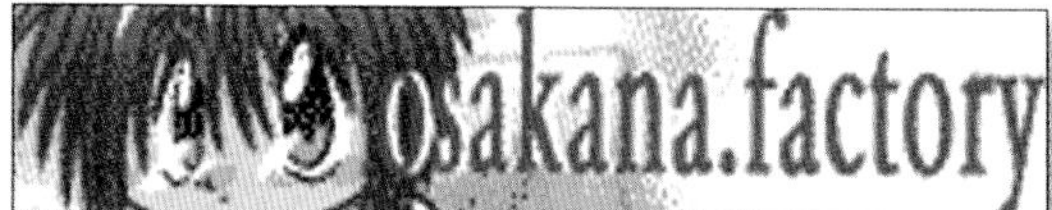

http://www.sfc.keio.ac.jp/~t96639yn/mishiki.html
- osakana.factory - / みしきさかな

http://www.netwave.or.jp/~margery/
Abracadabrante margery & MARGERY / margery

めいどいんぢゃぱん
http://www.asahi-net.or.jp/~nu7h-ootw/
眉毛

這個網站中，所登載的全都是相當符合網站名稱的MEIDO服飾，雖然其相當多元化，但不論哪一個作品都散發出MEIDO的風格。但是，MEIDO現在還有在創作嗎？

As the site name suggests, this home page features many images of characters wearing maid costumes. Although maid costumes are the theme of all the images, each is a unique design. By the way, I wonder if there are still maids like this in real life...

http://ha4.seikyou.ne.jp/home/akamaru/
Death Size / akamaru

http://home6.highway.ne.jp/teleusa/
きゃろっと美少女パラダイス / てれうさ

http://www.angel.ne.jp/~rinse/
DigitalGirl / SOARER

http://www.ne.jp/asahi/eifye/nps/
EIFYE原子力発電所 / EIFYE ABEL

ttp://www2q.biglobe.ne.jp/~ep-rom/rom/
EP-ROM Home Page / EP-ROM

http://www.os.rim.or.jp/~char/
G's club / GEDDY

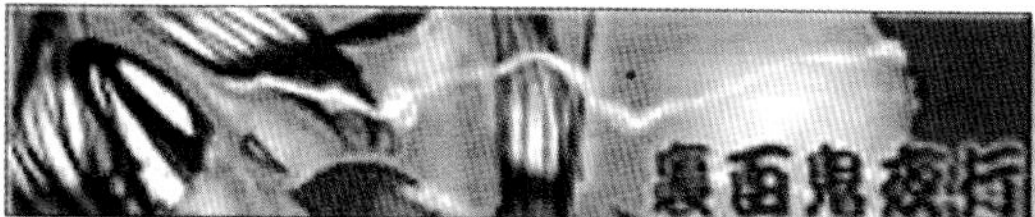

http://www.bekkoame.ne.jp/~mashitaka/
裏・百鬼夜行 / ましたか

http://home.att.ne.jp/gold/kenta/
WOODY HILL / 狼犬太

http://www.angel.ne.jp/~rinse/
おーだーメイド(closed) / SOARER

http://www.sfc.keio.ac.jp/~t96639yn/mishiki/overalls.html
オーバーオール友の会 / みしきさかな

http://www.din.or.jp/~onoe-clp/
ONOE's HomePage / ONOE

http://www.sic.shibaura-it.ac.jp/~l96073/
STUDIO S.D.T. / 結城辰也

http://www.campus.ne.jp/~suna/
SANDWORKS / 砂

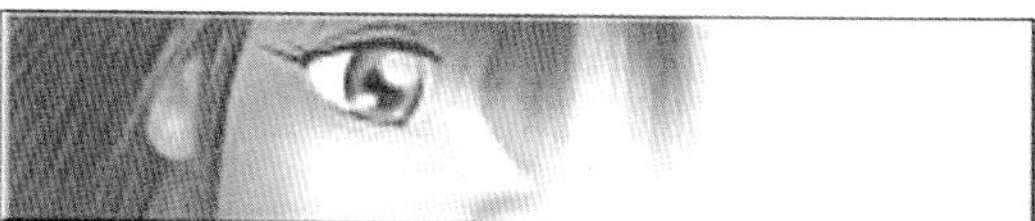

http://www.phoenix-c.or.jp/~yukino/
System町娘 / つるぎゆきの

http://ha4.seikyou.ne.jp/home/akamaru/

Death Size / akamaru

WOODY HILL

http://home.att.ne.jp/gold/kenta/

狼犬太

只要看服裝就知道這是狼犬太的網站。這裡登載的是戰隊單獨企劃的「水手兔女郎」。他們是企劃利用暗紅和暗粉紅，將美少女兔女郎化的惡組織「黑兔子」。

其他還有許多以美人魚、女侍等為題材的作品。

Kenta really likes girls in costume, it seems. In this book, images of two of his characters, "Dark Red" and "Dark Purple" from the "Dark Rabbit Syndicate" are shown. The syndicate is planning a "bishoujo" bunny conspiracy to make girls become bunnies. Besides the bunny images, there are pictures of mermaids, maids and waitresses.

http://www.netlaputa.ne.jp/~upiwo/
AquaMoon / うぴを

http://www.ddt.or.jp/~baibai/
ふらふら通り / ばいばい

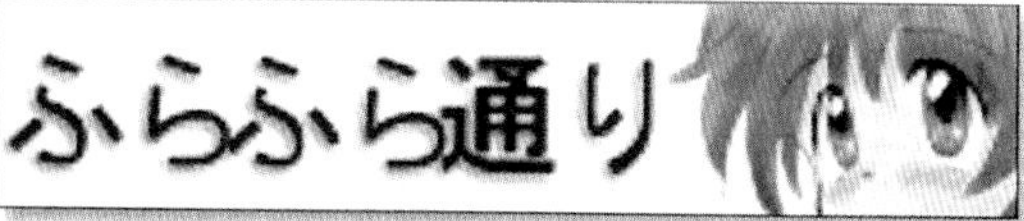

http://www.ddt.or.jp/~baibai/
ふらふら通り / ばいばい

http://member.nifty.ne.jp/KURUTO/
KURUTO's Museum / くると

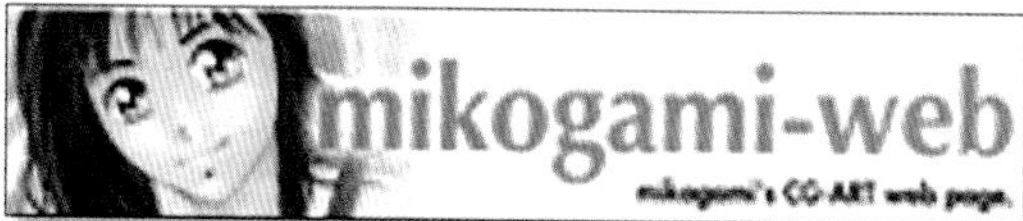

http://www.pluto.dti.ne.jp/~mikogami/
mikogami-web / 御琥神

http://w3.mtci.ne.jp/~myosi/
脳みそうにうに劇場 / 太田まさよし

http://www.remus.dti.ne.jp/~nyo/
Ny's Archives / 和泉如苑

http://www.remus.dti.ne.jp/~nyo/
Ny's Archives / 和泉如苑

http://www.remus.dti.ne.jp/~nyo/
Ny's Archives / 和泉如苑

http://www.remus.dti.ne.jp/~nyo/
Ny's Archives / 和泉如苑

http://w3.mtci.ne.jp/~myosi/
脳みそうにうに劇場 / 太田まさよし

看了這個網站之後，才知道短的外罩衫又稱之為細褶罩衫，在脖子處車入鬆緊帶來做出細細的皺摺。我完全不知道，在洋裁用語中，sirring的意思就是「加上衣褶」。

This is the home page of Upiwo who is a big fan of the "Anna Miller's" restaurant chain in Japan. The waitresses of this restaurant are very well known and admired for their special uniforms. I learned from this page that the blouse which is worn by Anna Miller's waitresses is called a "shirring blouse".

http://www.mirai.or.jp/~hoehoe/
☆ACTIVISION☆ / ほえほえ

ttp://www.alles.or.jp/~terakata/
HAW PAR VILLA / 寺方和日子

http://www.tky.3web.ne.jp/~ellin/
Girls in Wonderland / 牧村えりん

http://plaza11.mbn.or.jp/~icecat/
氷猫は冬眠中 / 氷猫ICE

http://www.cc.rim.or.jp/~inkpot/
ねこみみ振興会 / いんくぽっと

http://www.mars.dti.ne.jp/~naitou/
嗚呼、我等加藤隼戦斗隊 / 加藤

http://plaza6.mbn.or.jp/~makura/
まくらの部屋 / 夢里まくら

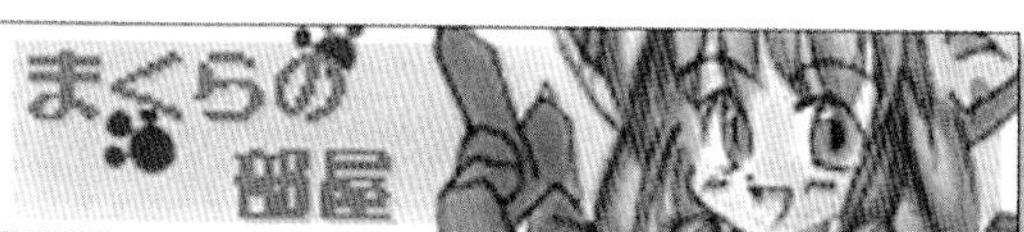

http://plaza6.mbn.or.jp/~makura/
まくらの部屋 / 夢里まくら

http://www.cc.rim.or.jp/~nekokan/
美紗魅 / 猫鑑¢

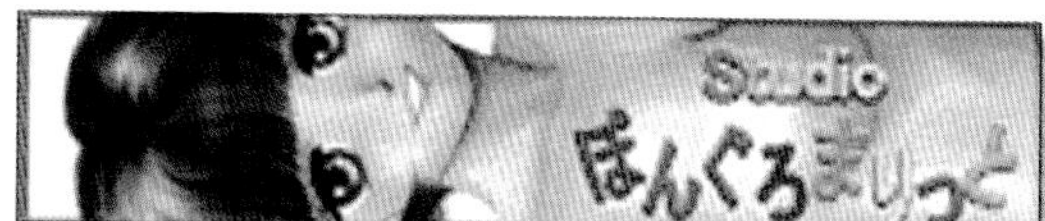

http://netpassport-wc.netpassport.or.jp/~wsonehar/
Studioぽんぐろまりっと / ぽんぐろ

http://netpassport-wc.netpassport.or.jp/~wsonehar/
Studioぽんぐろまりっと / ぽんぐろ

http://bacchus.brg.co.jp/tear/
nekodan / ていあ☆彡

嗚呼、我等加藤隼戦斗隊
http://www.mars.dti.ne.jp/~naitou/
加藤

這是可愛的野獸系列CG網站，作者加藤將喜愛動物的心情傳達給大眾，其在無意中融入了八字眉、近視眼的美少女，現在這個企劃CG正在籌劃中。

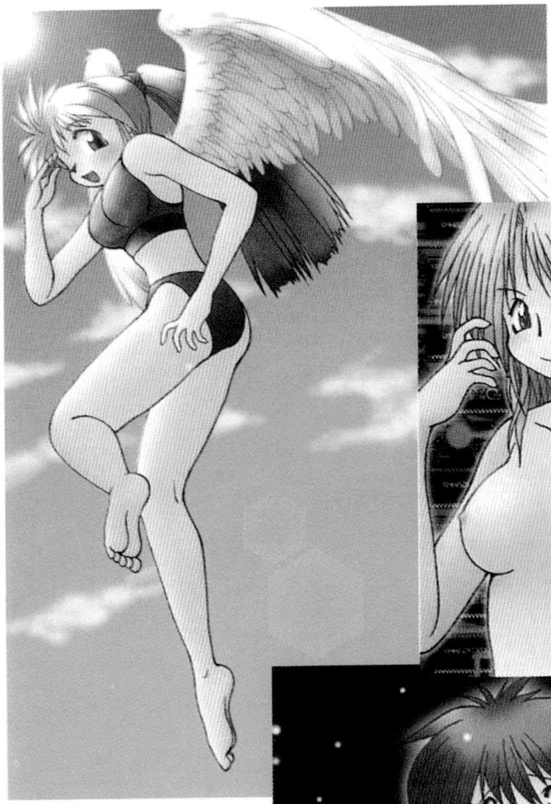

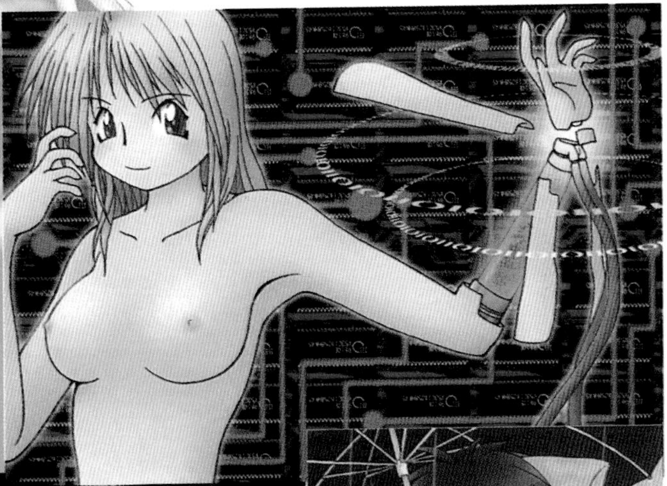

Cute Beast-like "kemono" girls are one of Kato-san's specialties. After browsing through his pages you can guess he must really be a big animal-lover! Also here are images of girls with "tare-me" (eyes drawn in a cute down-turned style), "meido-san" (girls in maid outfits) and "megane-ko" (girls wearing glasses). A special section entitled "Monthly Project" treats visitors to a new seasonally themed image each month.

HQ's home page

http://www.cyborg.ne.jp/~hq_hp/index.html
HALF QUARTER's HP / 梶山弘

http://www.cyborg.ne.jp/~hq_hp/index.html
HALF QUARTER's HP / 梶山弘

http://www6.big.or.jp/~cat-brow/
Cat Brow Internet / サークルCat Brow

http://anya.org/fox/
きつね友の会公民館 / delica

http://home.interlink.or.jp/~oyaman/
IMAGE FACTORY / 尾山泰永

http://plaza3.mbn.or.jp/~kona/
KEYWEST / KONA

http://www.luvnet.com/peoples/maruto/
泥酔桜国 / まると!

http://www.luvnet.com/peoples/maruto/
泥酔桜国 / まると!

http://www.geocities.co.jp/Playtown-Denei/8930/Homepage.htm
みぽずシステム【MIF】/ 黒猫

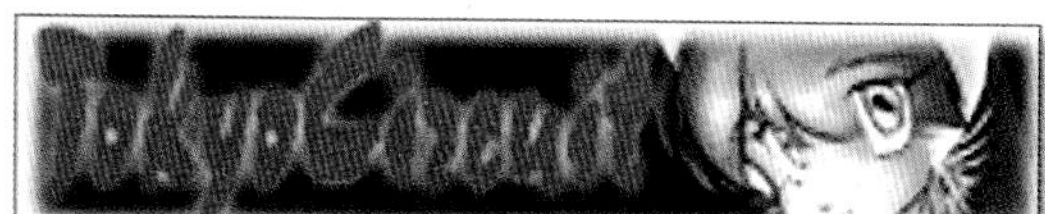

http://anya.org/t_circuit/
Tokyoサーキット / delica

http://anya.org/t_circuit/
Tokyoサーキット / delica

http://www.fsinet.or.jp/~jfujita/
J.Fujitaの納戸 / 藤田　純也

泥酔桜国
http://www.luvnet.com/peoples/maruto/
まると！

作者MARUTO於漫畫、插畫等商業雜誌上非常活躍。他描繪的野獸美女非常可愛，圓圓的臉頰以及色彩鮮豔的自然配色，這些都是他的特色。網站中也有英文網頁

Maruto!, the author of this home page, is very active as a cartoonist and illustrator and is always on the lookout for new projects. Maruto! specializes in "kemono" girls which are extremely cute with "Lori" or Lolitaish-looking faces and bright colors. He also has pages in English.

http://www.ade.internet.ne.jp/
ADE official homepage / ARTDESIGN

http://www02.u-page.so-net.ne.jp/gb3/b-cat/
AMETHYST Jewel BOX / くろねこ大王

http://www02.u-page.so-net.ne.jp/gb3/b-cat/
AMETHYST Jewel BOX / くろねこ大王

http://www02.u-page.so-net.ne.jp/gb3/b-cat/
AMETHYST Jewel BOX / くろねこ大王

http://www.na.rim.or.jp/~fre/anan/
友遊広場 / ANNANDULE Project

http://www.bekkoame.ne.jp/~ansoft/
ANSoft HOME PAGE / ANSoft社長　永田氏

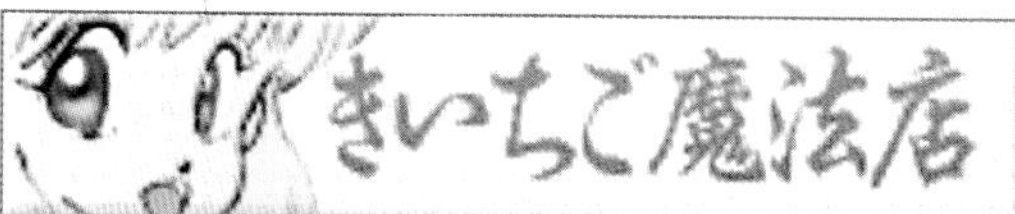

http://www.vector.co.jp/authors/VA004239/
きいちご魔法店 / AOI☆

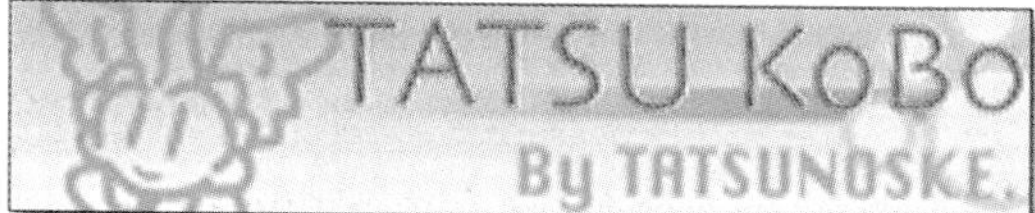

http://www.vector.co.jp/authors/VA012144/
TATSUKobo / たつのすけ

http://plaza7.mbn.or.jp/~secret_club/
秘密倶楽部ほぉ～むぺぇ～じ / Syun&星河☆苗

http://www.alles.or.jp/~pyon/
Carrot House / pyon

http://www.vector.co.jp/authors/VA006860/
爆裂健 Home Page II / 爆裂健

http://www1.odn.ne.jp/~aaa33460/
爆裂健 Home Page I / 爆裂健

http://karakuri.com/
美少女動画同人ソフト C.A.T. / C.A.T.

http://www.lares.dti.ne.jp/~crowd/
CROWD HomePage / CROWD

http://www.vector.co.jp/authors/VA004239/
きいちご魔法店 / AOI☆

http://ait.u-aizu.ac.jp/~s1042063/
F.E.C.soft HomePage / Azuman

http://ha2.seikyou.ne.jp/home/harumaki/
スタジオ インパクト / スタジオ インパクト

http://www.tky.3web.ne.jp/~nanashi/
Nanashi-soft / 岸本勝司

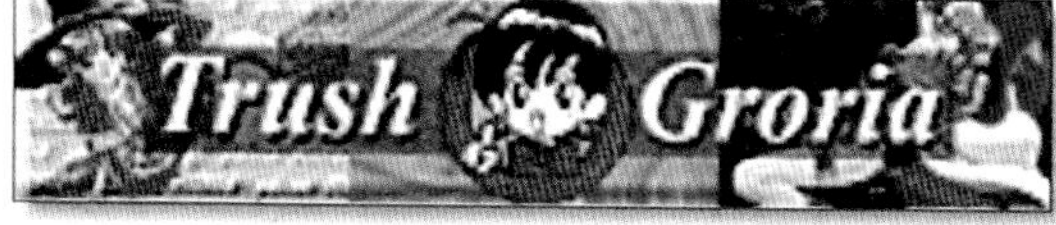

http://www.infonia.ne.jp/~fix/
Trush&Groria's Home Page / Trush

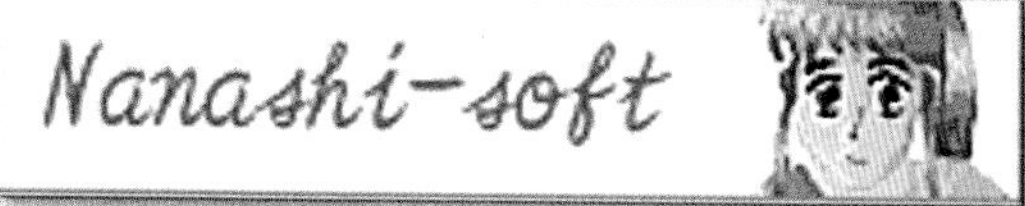

http://www.tky.3web.ne.jp/~nanashi/
Nanashi-soft / 岸本勝司

ADE official homepage
http://www.ade.internet.ne.jp/
ARTDESIGN

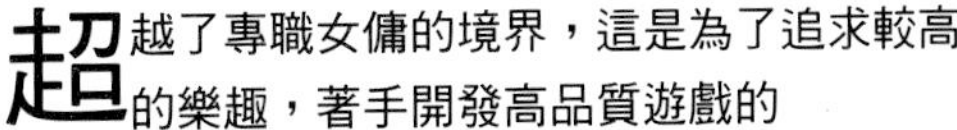

超越了專職女傭的境界，這是為了追求較高的樂趣，著手開發高品質遊戲的ARTDESIGN ENTERTAINMENT（以下稱ADE）的正式網頁。他以創造者之姿，支援個人提供他們活動的場所。
登載於這個網站的作品是ADE的幻想創作。

This is the official home page of ARTDESIGN ENTERTAINMENT (ADE), a company formed by a group of game creators who investigate new and original ideas for creating high-quality and interesting games. ADE supports creators by giving them a place to present their artwork. The girl character shown here is Minto-chan, ADE's mascot.

http://www.bekkoame.ne.jp/ha/takumi/
Bloody Moon / 秋月だんご

http://www.bekkoame.ne.jp/ha/bubu/
BUBU漬 / BUBU

http://plaza21.mbn.or.jp/~close/
Closed World / 黒川 いずみ

http://plaza21.mbn.or.jp/~close/
Closed World / 黒川 いずみ

http://www.st.rim.or.jp/~ctama/
CTAMA's HomePage / Cたま

http://www.neti.ne.jp/~ctama/
CTAMA's HomePage SideB / Cたま

http://www.bekkoame.ne.jp/ro/sion/
DARK ROOM / 紫苑

http://www.cc.rim.or.jp/~lyma/
DRAGON BLOOD / たいらはじめ

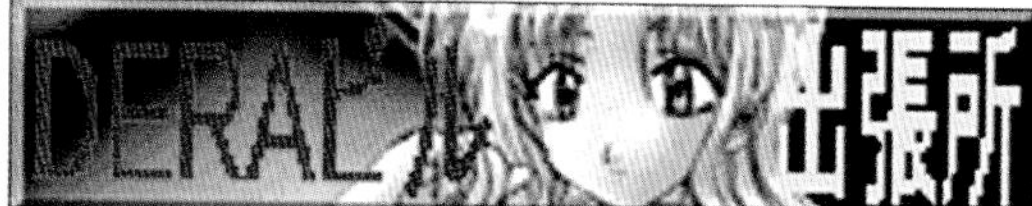

http://www3.alpha-net.or.jp/users/dera/
DERAビル出張所 / DERA

http://www.bekkoame.ne.jp/~hfhori/
Dynamic Lolita Library / ほりもとあきら

http://www.bekkoame.ne.jp/i/ga2598/
GODy the MAC! / ゴディ

http://www.bekkoame.ne.jp/i/hannaya/
はんな屋 / はりけんはんな

http://www.bekkoame.ne.jp/ro/ha15665/
My Wind / official doll

http://www.urban.ne.jp/home/nao9173/
a leisured parson N / NAOKI N

http://www.bekkoame.ne.jp/i/safire/
SAFIRE / SAFIRE

http://www.bekkoame.ne.jp/~toten/
TOTEN's ROOM / トーテンコブ

BUBU GALLERY 漬

BUBU漬
http://www.bekkoame.ne.jp/ha/bubu/
BUBU

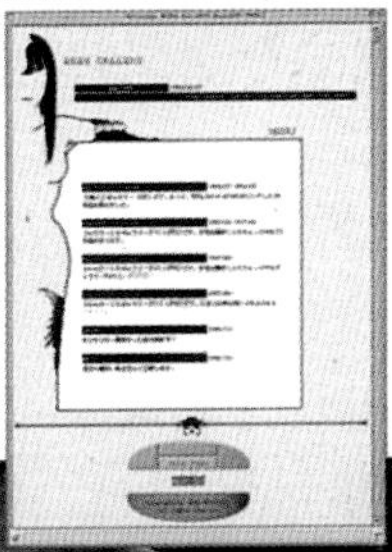

BUBU網站有著設計精美的網頁，網頁本身就相當藝術，往返於其網頁中讓人心曠神怡，雖然BUBU的作品是屬於色情作品，但他的作品也已達到藝術領域。

Bubu's home page design is very beautiful. The pages themselves are like pieces of art. Browsing page by page I really had an enjoyable experience. Bubu's images have an erotic feel and have truly reached the level of high art.

ADULT

http://www.tt.rim.or.jp/~aoao/
BLUE MAGIC / 青

http://www02.so-net.ne.jp/~dam/
妄想画廊 / 高橋ダム

http://www.ceres.dti.ne.jp/~mine-/
地雷屋 / MINE

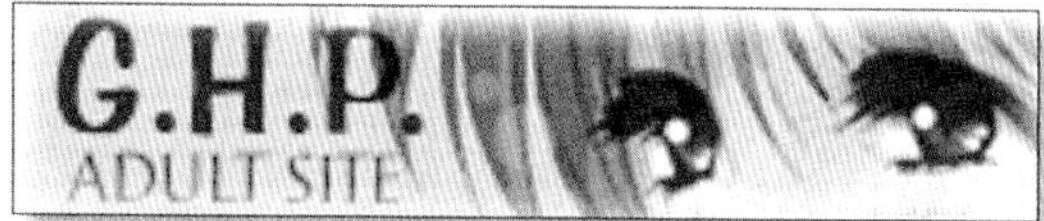

http://www.sh.rim.or.jp/~cke/
G.H.P. / c-ke

http://www.sh.rim.or.jp/~cke/
G.H.P. / c-ke

http://www2s.biglobe.ne.jp/~osa_hg/
DIGITAL ODDS / 大沢弾

http://www.lares.dti.ne.jp/~u-tom/
Gallery of Dreams / U-TOM

http://www.t3.rim.or.jp/~isapi/
Blue Cellar / いさぴ

http://www.din.or.jp/~jitta/
comic Jitta / John.J.Binbo

http://www.ipc-tokai.or.jp/~hyde/
PoisonArts 致命傷/FATALISM WORKS / 弥舞秀人

http://www.tk.airnet.ne.jp/acty/ponk/
ponk The Funkey Bunney / 梅田"ACTY"大丸

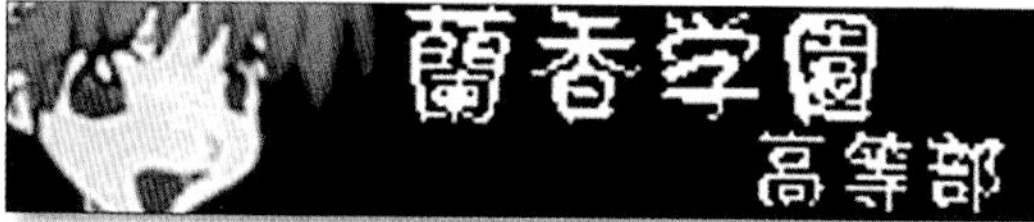

http://w33.mtci.ne.jp/~s707/
蘭香学園高等部 / 人造人間s707号

http://w33.mtci.ne.jp/~s707/
蘭香学園高等部 / 人造人間s707号

http://w33.mtci.ne.jp/~s707/
蘭香学園高等部 / 人造人間s707号

http://www.din.or.jp/~aobun/gate0.html
サディスティック・アルケミィ～嗜虐的錬金術～ / 青文鳥

http://www1.akira.ne.jp/~shoma/
山野草 / 雪原 露

地雷屋
http://www.ceres.dti.ne.jp/~mine-/
MINE

登載著MINE眉飛色舞的CG作品，這個網站中有禁止十八歲以下觀賞的網頁，因為設計了相當困難的“口令”，所以並不是那麼容易進入的。

Illustration/MINE

The expressively arched eyebrows and eyes of the characters shown here are very sexy-looking. This site includes pages which are forbidden to those under age 18 but these pages are strictly password protected.

ADULT

http://www.age.ne.jp/x/kiske/
第七南瓜部隊 / KISKE

http://www.bekkoame.ne.jp/i/gf8160/
DEEP CAVE / ピエールのらの

http://www.bekkoame.ne.jp/~beldandy/HolyBell/
ほぉりぃ☆べるのお絵カキの〜と / ほぉりぃ☆べる

http://www.bekkoame.ne.jp/~snake-pit/
ちぬちぬ少女の王国 / SNAKE-PIT

http://plaza24.mbn.or.jp/~watermelon/
カフェテリアWATERMELON / kosuge

http://www.cosmos.co.jp/~ramao/
らpcon / くま坂らま男

http://plaza28.mbn.or.jp/~miragenovels/
幻影図書館 / ミラージュ

http://plaza19.mbn.or.jp/~kojie/USF/
Underground Sample Files / 孤児郎

http://plaza19.mbn.or.jp/~kojie/USF/
Underground Sample Files / 孤児郎

http://plaza13.mbn.or.jp/~studiovanguard/
STUDIO VANGUARD / 安童あづ美

DEEP CAVE

http://www.bekkoame.ne.jp/i/gf8160/

ピエールのらの

從有點可疑又棒呆了的網頁進入這個網站，其登載著獨特風格的CG，以及以格鬥遊戲為題材的作品，這是充滿個人風格的網站。

Starting from the rather mysterious and cool top page to the gallery section, you can see original character illustrations and fighting game characters. You can really feel the author's passion for self-expression through his work.

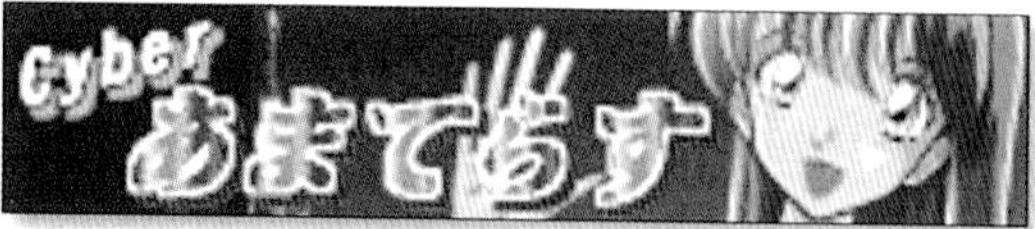

http://www.fureai.or.jp/~kumagon/
Cyberあまてらす / 熊木十志和

http://www.din.or.jp/~rinn/bnnr/
ばな〜倶楽部 / Woody-RINN

http://www.din.or.jp/~rinn/bnnr/
ばな〜倶楽部 / Woody-RINN

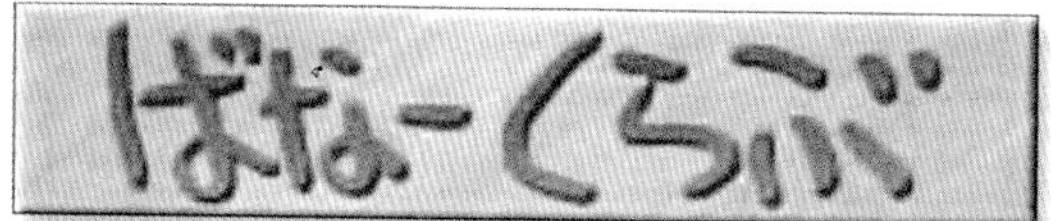

http://www.din.or.jp/~rinn/bnnr/
ばな〜倶楽部 / Woody-RINN

http://www.ask.or.jp/~mtint/fsn/
幻想の星空 / 翔空聖流

http://member.nifty.ne.jp/HeavensGate/
Heavens' Gate / 週休 8 日計画

http://member.nifty.ne.jp/HeavensGate/
Heavens' Gate / 週休 8 日計画

http://member.nifty.ne.jp/HeavensGate/
Heavens' Gate / 週休 8 日計画

http://member.nifty.ne.jp/HeavensGate/
Heavens' Gate / 週休 8 日計画

http://member.nifty.ne.jp/HeavensGate/
Heavens' Gate / 週休 8 日計画

http://www.catnet.or.jp/kanisan/net/
かにさんネット / かにさんネット

http://www.catnet.or.jp/kanisan/net/
かにさんネット / かにさんネット

http://www.catnet.or.jp/kanisan/net/
かにさんネット / かにさんネット

http://magmag.kaynet.or.jp/
まぐまぐパラダイス / まぐまぐソフト

http://magmag.kaynet.or.jp/
まぐまぐパラダイス / まぐまぐソフト

http://home.interlink.or.jp/~hinata/namifuku/
ナミフクDM / ひなた☆すう

http://w32.mtci.ne.jp/~noa_com/index.htm
NOA Company / Yoshi

http://w32.mtci.ne.jp/~noa_com/index.htm
NOA Company / Yoshi

http://w32.mtci.ne.jp/~noa_com/index.htm
NOA Company / Yoshi

http://www.asahi-net.or.jp/~fv3n-wkby/top.html
ぱやしのページ / ぱやし

幻想の星空 FantasticStarNight

幻想の星空
http://www.ask.or.jp/~mtint/fsn/
翔空聖流

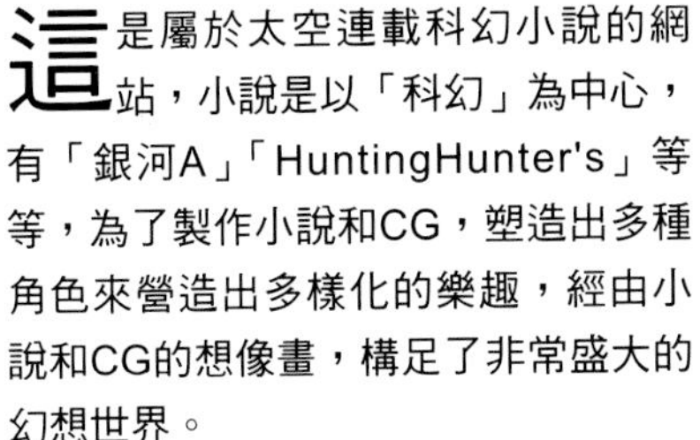

這是屬於太空連載科幻小說的網站，小說是以「科幻」為中心，有「銀河A」「HuntingHunter's」等等，為了製作小說和CG，塑造出多種角色來營造出多樣化的樂趣，經由小說和CG的想像畫，構足了非常盛大的幻想世界。

This site features fantasy serial novels. The novels you can read here are entitled, "Illusion", "Galaxy A" and "Hunting Hunters". The novels and CG images on this site are designed by various people so you can enjoy the individual tastes of each creator. The combination of these authors' novels and CG lead you to experience an enormous universe.

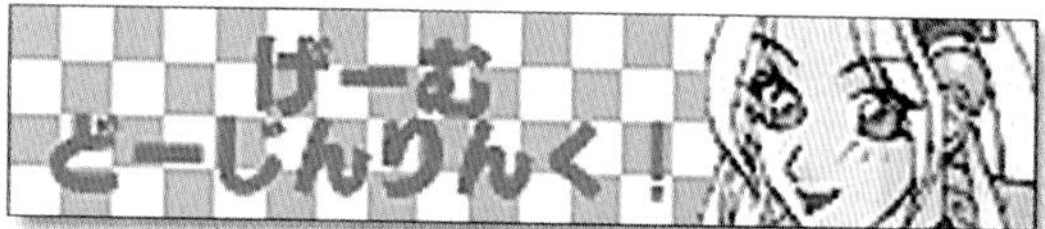

http://www.ohkini.net/~takeori/gdl/
げーむどーじんりんく / 武折貴子

http://www.ohkini.net/~takeori/gdl/
げーむどーじんりんく / 武折貴子

http://www.ohkini.net/~takeori/gdl/
げーむどーじんりんく / 武折貴子

http://www.halcyon.ne.jp/
Net HALCYON on World Wide Web / Net HALCYON

http://www6.big.or.jp/~kimuti/
KIMUTIのひみつのお部屋 / KIMUTI

http://www02.so-net.ne.jp/~nissan/
MAT-NET HomePage / NISSAN( )

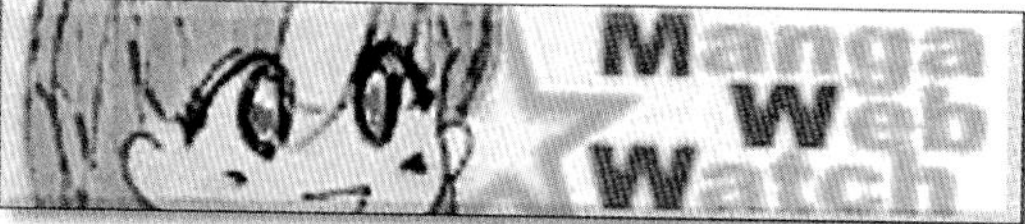

http://www.campus.ne.jp/~kim/manga.html
MangaWebWatch / 木村

http://www2n.biglobe.ne.jp/~syoujyor/
ランジェリンク / 398

http://sa.sakura.ne.jp/~syoujyor/
るりんく / 398

http://sakuya1.izayoi398.toshima.tokyo.jp/~s398398/index.htm
KINGS FEILD / 398

http://www.kulawanka.ne.jp/~takahiro/skill/index.html
SKILL SEEKERS / TAKAHIRO

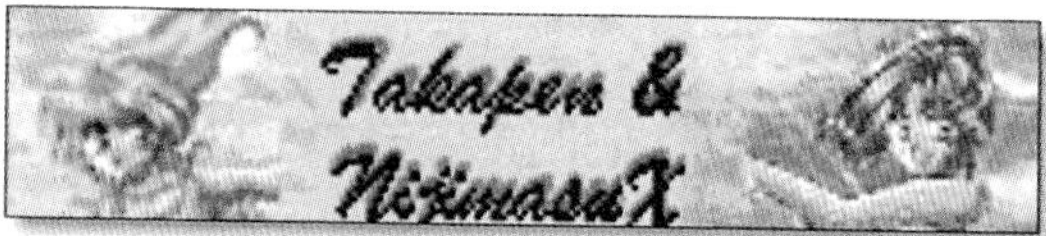

http://w33.mtci.ne.jp/~takapen/
Takapen & NijimasuX's HomePage / Takapen

http://w33.mtci.ne.jp/~takapen/
Takapen & NijimasuX's HomePage / Takapen

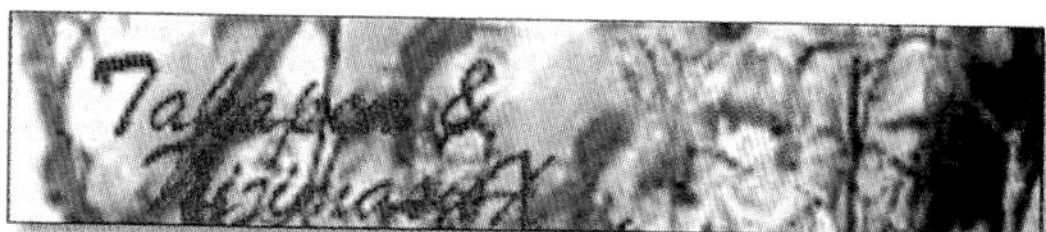
http://w33.mtci.ne.jp/~takapen/
Takapen & NijimasuX's HomePage / Takapen

http://www.raidway.ne.jp/~iharay/
うるさい小娘 / とりのすけ

http://www.raidway.ne.jp/~iharay/
うるさい小娘 / とりのすけ

http://www1.odn.ne.jp/~aaa04690/
100人規模 全国展開サークル「うれら」 / うれら

投稿

感謝多方的協助來完成這本書，這個網頁登載著其他的投搞作品。

http://user.shikoku.ne.jp/hmx12mlt/
風の森工房 / マサ

http://ww1.tiki.ne.jp/~su-sky/
Lost Memory/す～

http://home.intercity.or.jp/users/K/
KENホームページ / KEN

http://home.intercity.or.jp/users/K/
KENホームページ / KEN

http://ww1.tiki.ne.jp/~su-sky/
Lost Memory/す～

http://www.geocities.co.jp/Playtown/3603/
KOCHAN.STUDIO ジオシティーズ営業所
KOCHAN

http://www.lares.dti.ne.jp/~hazeou/
KARASAGI / 立花はぜお

Here are briefly introduced several homepages from which various artwork was contributed. I appreciate everyone's cooperation.

http://www.na.rim.or.jp/~nishino/
Asymmetric Inclination Society / にしのりん

http://www.raidway.or.jp/~hiroki_k/
Apple Town / 緋色樹

http://www.raidway.or.jp/~hiroki_k/
Apple Town / 緋色樹

http://www.huhka.com/
HUHKA-DO / ふーか

http://www.huhka.com/
HUHKA-DO / ふーか

http://www.mirai.ne.jp/~gotop/
Peeping P! / ごとP

http://www.mars.dti.ne.jp/~hot/
HOT SEAT - 物延法度のおもちゃ箱 / 物延法度

http://www.sm.rim.or.jp/~nmr/GAME/magic/
目んたまTシャツ王国 / NMR

http://www.fitweb.or.jp/~nyao/
studio necohouse / ねこのにゃお

http://www.fitweb.or.jp/~nyao/
studio necohouse / ねこのにゃお

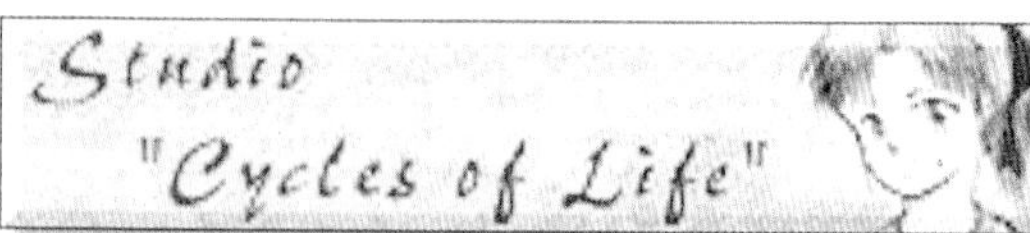

http://www.sm.rim.or.jp/~nmr/
Studio Cycles of Life / NMR

http://www.cyborg.ne.jp/~kana/
Tajima Show Room / たじまはるみ

http://www2.justnet.ne.jp/~ramiki/twilight/
Twilight Aquarium / 川羽らみき

http://www.mars.dti.ne.jp/~dggc/
UpperD3 On Line / kimtoy

http://www.nue.net/~wahyomi/
Vortex of Arts. / 水谷わひょみ

http://www.phoenix-c.or.jp/~e-db8/
negative-doll / D.D

studio necohouse
http://www.fitweb.or.jp/~nyao/
ねこのにゃお

網站中多數是屬於貓耳風格的作品，作品種類繁雜。

Thoughout this site you can enjoy CG of "neko-mimi" (cat ear) "kemono" characters. This site also includes various other images however. The Fei Yen game character image drawn in Neko-no-nyao's own special style is very cute and one of my favorites.

INDEX

●左から、サイト名・作者名・URL・ページ。掲載はコード順。太数字は、CG作品掲載ページを示す。

サイト名	作者名	URL	ページ
[仮庵]	島田朝臣	http://member.nifty.ne.jp/asomi/	30・**31**
- osakana.factory -	みしきさかな	http://www.sfc.keio.ac.jp/~t96639yn/mishiki.html	100
-R C A-	Endless	http://www.tk.xaxon.ne.jp/~endless/	72
-番長列伝- MIG-29's Home Page	MIG-29	http://www.synapse.ne.jp/~bancho/	98
<蔀>しとみ	RYOTA	http://www.ktroad.ne.jp/~ryota/	10
☆ACTIVISION☆	ほえほえ	http://www.mirai.or.jp/~hoehoe/	106
○○FANのほ～むぺ～じ!	アニック	http://village.infoweb.ne.jp/~fan/	70
○急電鉄	きよ○	http://www.geisya.or.jp/~kiyo/	100
100人規模 全国展開サークル「うれら」	うれら	http://www1.odn.ne.jp/~aaa04690/	120
1inch	朝妻天	http://www.w-w.ne.jp/~1inch/	28・**29**
21世紀計画社・電脳社内報	21世紀計画社	http://www.yo.rim.or.jp/~takeori/21th/	72
2B鉛筆工房	芥	http://www.nsknet.or.jp/~akuta/	58
3DCG補完委員会	WAKA	http://www.asahi-net.or.jp/~fv3n-wkby/	98
3G	Joe	http://www.win.ne.jp/~tomochan/3g.html	98
42番街のペテン師たち	Lips	http://www.asahi-net.or.jp/~pg4s-nw/	72
a leisured parson N	NAOKI N	http://www.urban.ne.jp/home/nao9173/	112
A RAILROAD JUNCTION	T.Kawasaki&YOU	http://www.jah.ne.jp/~next96/	36
A tender blue breeze	あめいすめる	http://www.webnik.ne.jp/~uzumi/	78・**79**
A.S.PANORAMIC	佐藤明機	http://www2.gol.com/users/batcave/	60
Abracadabrante margery&MARGERY	margery	http://www.netwave.or.jp/~margery/	**100**
ACHARIN's WORLD	ACHARIN	http://www2a.biglobe.ne.jp/~acharin/	54
ADE official homepage	ARTDESIGN	http://www.ade.internet.ne.jp/	110・**111**
Aggregat WORKS	悠瀬巧一	http://www01.u-page.so-net.ne.jp/ja2/ko_yuuse/	32
Ah Mitsuketa!	高橋哲人	http://www.bekkoame.ne.jp/~tettete/	22
Algernon's Artless Arts	Algernon	http://www.din.or.jp/~alger/	12
Alisa The Wonderland	Mokkun	http://www.cc.rim.or.jp/~mokkun/	10
AMETHYST Jewel BOX	くろねこ大王	http://www02.u-page.so-net.ne.jp/gb3/b-cat/	110
ANSoft HOME PAGE	ANSoft社長　永田氏	http://www.bekkoame.ne.jp/~ansoft/	110
Anze's HOMEPAGE OasisRoad	杏世	http://www.sainet.or.jp/~anze/	32・**33**
AOI (E) UNIT	緒空賀葵	http://www.bekkoame.ne.jp/~a_eu/	92
Apple Town	緋色樹	http://www.raidway.or.jp/~hiroki_k/	122
APPLICANT	セイン流	http://www02.so-net.ne.jp/~ryu-sein/	32
AquaMoon	うぴを	http://www.netlaputa.ne.jp/~upiwo/	104・**105**
ARTCH	宮村和生	http://www.din.or.jp/~abara/	24
ASTRO LINER	pukuten	http://www.webnik.ne.jp/~pukuten/	32
ASYLUM (アサイラム)	I M I	http://village.infoweb.ne.jp/~imi/	72
Asymmetric Inclination Society	にしのりん	http://www.na.rim.or.jp/~nishino/	122
atelier-Suite Grove-	睡宵堂	http://member.nifty.ne.jp/suishoudo/	**48**

AXIS WORLD	KOU	http://www.yk.rim.or.jp/~axis88/	76
B-PART	たちばなふゆか	http://www.bekkoame.ne.jp/i/gd5293/	74
BADCOMPANY	恋川春町	http://member.nifty.ne.jp/BADCOMPANY/	64
Bahamut's Garden	Bahamut	http://www.fang.or.jp/~bahamut/	94・**95**
BIG BEN	N.I	http://www-user.interq.or.jp/~naoki/	32
bit of GLASS FOREST	里見桜次	http://www.ceres.dti.ne.jp/~snowdrop/	40
BLACK BOX	柴崎泪	http://www.netlaputa.ne.jp/~louie/	76
BLACK STARS	みうらたけひろ	http://www.annie.ne.jp/~miura/	18
Bliss and Rapture...	瑞穂	http://home4.highway.ne.jp/mizuho/how/	24
Bloody Moon	秋月だんご	http://www.bekkoame.ne.jp/ha/takumi/	112
BLUE BLACK	葉月	http://www.tokyo-bay.or.jp/~yamane/	86
Blue Cellar	いさぴ	http://www.t3.rim.or.jp/~isapi/	114
BLUE MAGIC	青	http://www.tt.rim.or.jp/~aoao/	114
Brain NoiZ	こば	http://www.lofty-tec.co.jp/~koba/	32
brilliant street	bitter	http://www.kisnet.or.jp/bitter/	32
BUBU漬	BUBU	http://www.bekkoame.ne.jp/ha/bubu/	**112・113**
CACTUS	酒田みりん	http://www.rr.iij4u.or.jp/~mirin/	78
Cafe pasta	榊原薫奈緒子	http://www.tsp.ne.jp/~s_naoko/	32
Cafe☆ミぽっとはうす	葉桜みずな	http://w33.mtci.ne.jp/~mizuna/	30
Candy Box	クリスト・マーク	http://www2k.biglobe.ne.jp/~candybox/	32
Carrot House	pyon	http://www.alles.or.jp/~pyon/	110
Cat Brow Internet	サークルCat Brow	http://www6.big.or.jp/~cat-brow/	108
Cat's Face	水花	http://member.nifty.ne.jp/CATSFACE/	50・**51**
Cat's Labo	ねこやま工房	http://www.osk.3web.ne.jp/~catslabo/	42
CG Time	まるたか	http://www2s.biglobe.ne.jp/~Marutaka/	34
CHARMING SQUARE GARDEN	CHARM	http://www.os.rim.or.jp/~charm/	58
Cherry Pie Web	さくらぎけい	http://www.vector.co.jp/authors/VA004639/	100
CHILDMOON	つくみや	http://www2.mwnet.or.jp/~tnaoko/	32
CHI倶楽部	タカハマ	http://www.ic-net.or.jp/home/chic/	78
CHOCOLATE CHIFFON	大阪なる	http://www2s.biglobe.ne.jp/~chiffon/	76
Chocolate Paradise	GPX-TAKA	http://www2.osk.3web.ne.jp/~gpxtaka/	40
Circle RGB	RGB	http://www.tky.3web.ne.jp/~rgb/	76
Closed World	黒川 いずみ	http://plaza21.mbn.or.jp/~close/	112
comic Jitta	John.J.Binbo	http://www.din.or.jp/~jitta/	114
CornPine	SOTOKAN	http://member.nifty.ne.jp/sotokan/	70
CPU's CG Gallery	CPU	http://www.sainet.or.jp/~cpunit/	40
Crazy Circus	楠たつき	http://www.netlaputa.ne.jp/~rei-t/tazki/	78
CROWD HomePage	CROWD	http://www.lares.dti.ne.jp/~crowd/	110
CROWN	さくら	http://www2s.biglobe.ne.jp/~crown/	32・34
CTAMA's HomePage SideB	Cたま	http://www.neti.ne.jp/~ctama/	112
cute core	早坂奈槻	http://www.os.urban.ne.jp/home/hayasaka/	86

Cyberあまてらす	熊木十志和	http://www.fureai.or.jp/~kumagon/	118
D.P.I.	Sat	http://www.annie.ne.jp/~sat/	60
D・RA・GO・O・N!いんたーねっと	結維	http://www.kt.rim.or.jp/~youie/	96
D-SeaNET	sachi	http://plaza11.mbn.or.jp/~Sekiya/dsean/	10
DAP-ENTERPRISE	だっぷ	http://www.netlaputa.ne.jp/~dap/	60
DARK ROOM	紫苑	http://www.bekkoame.ne.jp/ro/sion/	112
DayDream	碧竜ヲサム	http://www2.marinet.or.jp/~sakuraba/	96
Death Size	akamaru	http://ha4.seikyou.ne.jp/home/akamaru/	**102**
Deep	シン・オオバ	http://www2.odn.ne.jp/~cae99980/	34
DEEP CAVE	ピエールのらの	http://www.bekkoame.ne.jp/i/gf8160/	116・**117**
Delusions of grandeus	MON	http://www.kt.rim.or.jp/~hymz/	60
DERAビル出張所	DERA	http://www3.alpha-net.or.jp/users/dera/	112
DIGITAL ODDS	大沢弾	http://www2s.biglobe.ne.jp/~osa_hg/	114
Digital 光画堂	二條ゆたか	http://member.nifty.ne.jp/yutaka_n/	30
DigitalGirl	SOARER	http://www.angel.ne.jp/~rinse/	102
DOIchan! HomeStadium	DOIchan!	http://www.win.ne.jp/~doichan/	34
DOUBLE TAP	榛名まお	http://www.tk.xaxon.ne.jp/~mao_h/	**56**
Downstairs Sweetheart - 階下の恋人 -	さくらんぼ	http://www.elf.ne.jp/~aki/	**54**・**55**
DPM-Page	土方幹	http://www.remus.dti.ne.jp/~dpm/	34・**35**
DRAGON BLOOD	たいらはじめ	http://www.cc.rim.or.jp/~lyma/	112
DRAGON PALACE	原田竜介	http://www.ceres.dti.ne.jp/~yharada/	76
Drawing Factor	OCHA	http://www.cc.rim.or.jp/~ocha/	82・**83**
Dynamic Lolita Library	ほりもとあきら	http://www.bekkoame.ne.jp/~hfhori/	112
E-KENIX	にんにん!	http://www.netlaputa.ne.jp/~kenix/	82
EARTH CG Gallery	EARTH	http://www2e.biglobe.ne.jp/~earth/	34
EDEN the DOORS official page	氷堂涼二	http://www.angel.ne.jp/~ryoji/	44
EIFYE原子力発電所	EIFYE ABEL	http://www.ne.jp/asahi/eifye/nps/	102
enocchy's ROOM	榎本 千春	http://home.interlink.or.jp/~chiharu/	20
EP-ROM Home Page	EP-ROM	http://www2q.biglobe.ne.jp/~ep-rom/rom/	102
Esel神聖帝国	Esel	http://www.kumagaya.or.jp/~koju/	96
Ever Green Ever Blue	源之助	http://www.din.or.jp/~sion/gen/	40
Eye Appeal	結稀	http://www.alles.or.jp/~knot/	60
F.E.C.soft HomePage	Azuman	http://ait.u-aizu.ac.jp/~s1042063/	110
FLAT PACKAGE	大東亜ゆう	http://www.try-net.or.jp/~yufu/	76
FLEETING HAPPINESS	綾波	http://www2.justnet.ne.jp/~aynm/	78
Forbidden Fruit	月末　円	http://home3.highway.ne.jp/magatu/	20
FOREST GALLERY	武原卓&しのぶ一樹	http://www.mfi.or.jp/takehara/	34
FORZA on the Web	寿圭祐	http://www.din.or.jp/~kotobuki/	62
Forza!	のぶまる	http://member.nifty.ne.jp/marui/	100
Fragment Time	泰大	http://www.scan-net.or.jp/user/taidai/	36
FRE's お絵かきページ	FRE	http://www.na.rim.or.jp/~fre/	34

FREEZE MOON	Take-Four	http://www.alles.or.jp/~takefour/	12
FROM 雪待月	佐久間 和弥	http://www2s.biglobe.ne.jp/~fromy/	56
FUN	不能花	http://www.hh.iij4u.or.jp/~fun/	34
G to Z	G-メタル	http://www.yui.or.jp/~gmetal/	34
G.H.P.	c-ke	http://www.sh.rim.or.jp/~cke/	114
G's club	GEDDY	http://www.os.rim.or.jp/~char/	102
G3 Aim'sHomePage	Aim	http://www.alles.or.jp/~aim/	90
Gallery Capricious	よーすけ	http://plaza7.mbn.or.jp/~yosukef/	64・**65**
Gallery ILLUSION	BON	http://www3.big.or.jp/~bon/	96・**97**
Gallery of Dreams	U-TOM	http://www.lares.dti.ne.jp/~u-tom/	114
GENAX	GENAX	http://www2.justnet.ne.jp/~genax/	76
Girls in Wonderland	牧村えりん	http://www.tky.3web.ne.jp/~ellin/	106
gnosis net zero	Masaki Shibata	http://www.alles.or.jp/~morobosi/	30
Go-Q☆Factory	風渡カムイ	http://www2.tky.3web.ne.jp/~kamkam/	76
GODy the MAC!	ゴディ	http://www.bekkoame.ne.jp/i/ga2598/	112
GOLDEN BAT Home Page	GOLDEN BAT	http://www.threeweb.ad.jp/~bat6000/	34
GP工房	藤衣 悦巳	http://www.os.rim.or.jp/~etsumi-f/	76
Grass Valley	瑞杜秀智	http://nori.im.kindai.ac.jp/~myzz/	74
GRAVITATION	かずきゆい	http://www.bekkoame.ne.jp/~woodstok/yui/	14
GUIDE MAP	ほづみりや	http://www.bekkoame.ne.jp/i/gd5787/top.html	30
GUILD HOUSE	神之みわざ	http://www.ceres.dti.ne.jp/~miwaza/	10・**11**
H&T Laboratory	彪衣 亜騎(ひょうい あき)	http://www.infotrans.or.jp/~tabayan/	40
HALF QUARTER's HP	梶山弘	http://www.cyborg.ne.jp/~hq_hp/index.html	108
HAMIDASHI 情熱系	間枝皇二	http://www1.interq.or.jp/~d-hami/	76
HAT home page Heavy DE Show!!	HAT	http://www.kotonet.ne.jp/~tobo/	40・42
HAW PAR VILLA	寺方和日子	http://www.alles.or.jp/~terakata/	106
Heaven's Gate	SION	http://www.din.or.jp/~sion/	32
Heavens' Gate	週休8日計画	http://member.nifty.ne.jp/HeavensGate/	118
Henachocosystem On-Line	Project katz	http://www.tk.airnet.ne.jp/kotone_h/	82
HI-TECH CARROT	佳由樹	http://www.ruri.nadeshico.net/	78
HIGUCHIN [にこニコ DRAWINGS]	HIGUCHIN	http://home3.highway.ne.jp/higuchin/	92
HomePage PAI	PAI	http://www.sp.dianet.or.jp/~pai/	96
HOPE	はるの	http://www.campus.ne.jp/~takano	96
HOT SEAT - 物延法度のおもちゃ箱	物延法度	http://www.mars.dti.ne.jp/~hot/	122
HUHKA-DO	ふーか	http://www.huhka.com/	122
Hyoi's House Leaves	ひょい	http://club.pep.ne.jp/~hyoi.kaz/	18・**19**
ICE's CYBER ROOM	森永あいす	http://www.geocities.co.jp/SiliconValley/4718/	34
icenou	icenou	http://www2.odn.ne.jp/~caj16480/	98
ILLUST-GUERRILLA	NAKA	http://www.ddt.or.jp/~naka/	60
IMAGE FACTORY	尾山泰永	http://home.interlink.or.jp/~oyaman/	108
IMPERIAL PALACE	KWK36	http://www.yo.rim.or.jp/~kwk36/	40

IMPLACABLE	水羽輝幸	http://www2m.biglobe.ne.jp/~mizuha/	48
Impression of MY digital works	愛瀬雅巳	http://www.din.or.jp/~yitsuse/	58
In My Room	如月庵	http://home2.highway.or.jp/syui/	30
INNOCENT ARCADIA	瑞輝 智佳	http://www.md.xaxon.ne.jp/~arcadia/	34
INTERNET STUDIO AKI	三沢秋太郎	http://www.mars.dti.ne.jp/~aki-m/	94
IRIS営業所	IRIS	http://plaza11.mbn.or.jp/~iris/	72
iSOMEC's Home-Page	iSOMEC	http://www.sa.sakura.ne.jp/~isomec/	88・**89**
It's "NON" Parity!!	NON	http://www.win.ne.jp/~non/	58
It's a KATU'S HOMEPAGE!!	KATU	http://www.alles.or.jp/~easter/	90
ITAL GOODS Home Page	Ital Goods	http://hiei.okuma.nuee.nagoya-u.ac.jp/~itaru/	30
J.Fujitaの納戸	藤田　純也	http://www.fsinet.or.jp/~jfujita/	108
J.SAJI'S Homepage	佐治ジロー	http://www.big.or.jp/~hannaya/	66
JAPOTIO ATOLI	人身	http://www.pluto.dti.ne.jp/~hitom/	52
Jasmine Tea Break	Jasmine Tea	http://www.linkclub.or.jp/~shinno/	58
JEWEL MASTER	結城 梓	http://www2s.biglobe.ne.jp/~JEWEL/	20
JG-7	Ryu-Akt	http://www3.osk.3web.ne.jp/~yonehara/	82
Jiji's Home Page	じじ	http://www.ask.or.jp/~jiji/	78
JUNKBOX	萌沢ゆづる	http://www.bekkoame.ne.jp/i/yuduru/	76
K.P.G	Kokai	http://www.gld.mmtr.or.jp/~kokai/	80
K－いんぱくと	松莉 詠	http://home3.highway.ne.jp/matsuri/	66
Ka2DA Factory	Ka2DA	http://www.mars.dti.ne.jp/~ka2da/	82
kaedeはうす	楓 T2	http://www.bekkoame.ne.jp/i/kaede/	100
kamizanushi room	神座主志	http://www.dd.iij4u.or.jp/~kamiza/	30
KARASAGI	立花はぜお	http://www.lares.dti.ne.jp/~hazeou/	56・**121**
Karmic Relations!	成瀬ちさと	http://www.netlaputa.ne.jp/~naruse/	38・**39**
KAZUBOH Art Garden	KAZUBOH	http://members.xoom.com/okada/	92
Kei's Homepage	Kei	http://www.mars.dti.ne.jp/~kei/	34
KENホームページ	KEN	http://home.intercity.or.jp/users/K/	8・**9**・**121**
KEYWEST	KONA	http://plaza3.mbn.or.jp/~kona/	108
KIMUTIのひみつのお部屋	KIMUTI	http://www6.big.or.jp/~kimuti/	120
KINGS　FEILD	398	http://sakuya1.izayoi398.toshima.tokyo.jp/~s398398/index.htm	120
KITUPON'S　HP	KITUPON	http://jimcok.co.jp/lana/KITUPON/index.htm	72
KIYO'S CG ISLAND	KIYO	http://www.biwa.ne.jp/~k-t/	42
Knight of Dusk	落ち武者	http://www.din.or.jp/~otimusha/	**52**
KOCHAN STUDIO ジオシティーズ営業所	KOCHAN	http://www.geocities.co.jp/Playtown/3603/	86・**87**・**121**
Konomi Kuzuhara's Homepage	葛原このみ	http://www.mars.dti.ne.jp/~kuzuhara/	58
KURUTO's Museum	くると	http://member.nifty.ne.jp/KURUTO/	104
L・KAN研究所	うらかんRN	http://www.sun-inet.or.jp/~urakan/	82
LAZY VIRUS	たにざきひなり	http://www.din.or.jp/~acht/index.html	68
LEONPURPLE CG WORLD	れおんぱあぷる	http://www.sannet.ne.jp/userpage/leon/main.html	42
LEVEL	NaoXYZ	http://www.geocities.co.jp/Playtown/3331/	42

Life Like Love	石田あきら	http://www1.odn.ne.jp/ishida/	42
Lincoln Island	竜馬	http://www.remus.dti.ne.jp/~ryouma/	42
LINKS∇BOX ～黒猫館～	増多シイ夫	http://www.vega.or.jp/~shieo/	36
Lost Angel	みずき	http://www1.u-netsurf.or.jp/~mizuki/lost/	80
Lost Memory	す～	http://ww1.tiki.ne.jp/~su-sky/	68・70・**121**
LOVE SPIN DRIVE	おじゃわ	http://www.e-net.or.jp/user/totorin/	78
LSD ANNEX	おじゃわ	http://www.geocities.co.jp/Playtown-Denei/4872/	78
m/art-lab	Kyo.Komiya	http://jun.gaia.h.kyoto-u.ac.jp/~eiji/	72・**73**
MAGUROMEDIA	まぐろせんせい	http://www.nerimadors.or.jp/~maguro/	52
MangaWebWatch	木村	http://www.campus.ne.jp/~kim/manga.html	120
manna	朔実アンジ	http://www.sip.or.jp/~lockhart/	78
Marine Web	minoru	http://m-y.ed-sys.co.jp/	72
MarineMarilyn HomePage	みなかみ	http://www.linkclub.or.jp/~minakami/	52
Masaoki Satou's Homepage	佐藤正興	http://plaza8.mbn.or.jp/~masaoki/	66
MAT-NET HomePage	NISSAN( )	http://www02.so-net.ne.jp/~nissan/	120
megababes	emico and kilkeny	http://www1.sphere.ne.jp/emcweb/megababes/	**50**
Mei Itsuki's Web Page	伊月めい	http://www.blk.mmtr.or.jp/~itsuki/	28
MEI-Q-RONDO	珠梨やすゆき	http://www.din.or.jp/~syuriyan/	24
Memories	Coo	http://www.sannet.ne.jp/userpage/ko-ji/	**40**
meronの部屋	meron	http://plaza4.mbn.or.jp/~meron/	8
Metal Dress	犬飼 紅輝	http://www.alles.or.jp/~inukai/	88
MICKEY'S LAGOON	MICKEY	http://www.linkclub.or.jp/~mickey/	98
MIEKO'S PAGE	みえこ	http://home2.highway.or.jp/mmoriya/mieko/	56
mikogami-web	御琥神	http://www.pluto.dti.ne.jp/~mikogami/	104
MILK COCOA CAN	欧柊ここあ	http://www1.plala.or.jp/milk/	52
MILKY WATER	雅日あきら	http://www.os.rim.or.jp/~miyabie/	48
Mirage Field	レッド	http://www2s.biglobe.ne.jp/~n_kato/	44
MK2 FACTORY	めけめけ	http://www02.so-net.ne.jp/~mkmk/	82
MONA'S MONA	akeo	http://www.rnac.or.jp/~akeo/	44
Monotone Club'	かっちん	http://home3.highway.ne.jp/kattin/	16
Moonlight Aquarium	川原 美紀	http://nug.nasu-net.or.jp/~tmakino/	24
MOONRIYA	MOONRIYA	http://miya.or.jp/~moonriya/	72
MOTO'S CG ROOM	MOTO	http://www.kk.iij4u.or.jp/~moto/	82
MPLACABLE	水羽輝幸	http://www2m.biglobe.ne.jp/~mizuha/	48
mu genera	士門	http://www.bekkoame.ne.jp/~d_works/	92
Muhoho's Homepage!	むほほ	http://w32.mtci.ne.jp/~muho/	20
MUMU屋本舗	MUMU	http://www02.so-net.ne.jp/~mumu/	82
Mutation	みゅう	http://www.din.or.jp/~miu/	82
MUTIPETADOKYUN!!	かなこ	http://www.freepage.total.co.jp/hatuneandmulti/	82
my confidence world	上野真麻	http://www02.u-page.so-net.ne.jp/gb3/takizawa/	40
My Taste	Nori	http://www.sanynet.ne.jp/~nori/	54

My Wind	official doll	http://www.bekkoame.ne.jp/ro/ha15665/	112
Mystic Kingdom	こちこち	http://www.bekkoame.ne.jp/~kch/	16
Nanashi-soft	岸本勝司	http://www.tky.3web.ne.jp/~nanashi/	110
NBU 3DCG GALLERY	HIROPU	http://www.zokei.asu.ac.jp/~nbu/	98 · **99**
NB学園	或十せねか	http://www4.big.or.jp/~seneka/	38
Negative X	kasun	http://www2s.biglobe.ne.jp/~kas/	56
negative-doll	D. D	http://www.phoenix-c.or.jp/~e-db8/	38 · **122**
NegEdge	takoyaki	http://www.geocities.co.jp/Playtown/8106/	82
nekodan	てぃあ☆彡	http://bacchus.brg.co.jp/tear/	106
NEKOHOUSE	長居ねこ	http://www2.tky.3web.ne.jp/~nekomimi/	38
Net HALCYON on World Wide Web	Net HALCYON	http://www.halcyon.ne.jp/	120
NIGHT SHIFT	ういん	http://www.pluto.dti.ne.jp/~winn/	38
NishiTatsuki's Home Page!	にし☆たつき	http://www.asahi-net.or.jp/~kd4n-nsmt/stardust.htm	60
NIWAKEN's HomePage	NIWAKEN	http://www.mirai.ne.jp/~niwaken/	38
NOA Company	Yoshi	http://w32.mtci.ne.jp/~noa_com/index.htm	118
nobodyknows.	m_sugisaki	http://www.geocities.co.jp/Playtown-Denei/4194/	38
Nobu's BambooHouse	nobu	http://www.asahi-net.or.jp/~uw6n-tki/	72
Nocturne	ウララ	http://www.cc.rim.or.jp/~urara/	28
NOW WON!	NOW	http://www.ceres.dti.ne.jp/~n621197/index.htm	28
Ny's Archives	和泉如苑	http://www.remus.dti.ne.jp/~nyo/	104
Obsidian Wind JP	ive	http://www.opsdti.ne.jp/~ive/	94
Obsidian Wind USA	ive	http://mypage.goplay.com/obs/	94
OFF WORLD	フッタこーじ&ハッピーガヴァン	http://www.din.or.jp/~happy-g/	28
Oh!CULT Room	おかると	http://www.valley.ne.jp/~occult/	82
OKO HOME PAGE	oko	http://www.sam.hi-ho.ne.jp/~expo69/oko/	38
Olive Free State	高槻 悠	http://www.angel.ne.jp/~gazer/you/index.htm	54
One From The Heart	剣崎 星紅	http://home.interlink.or.jp/~okazakir/	66
ONOE's HomePage	ONOE	http://www.din.or.jp/~onoe-clp/	102
Open Your Eyes	海里	http://www.t3.rim.or.jp/~kairi/	52
ORIENT KOUYOU	荒葉	http://ha3.seikyou.ne.jp/home/kouyou/	22
P. H. S.（ぷに·ほえ·ステーション）	かかか	http://cgi3.osk.3web.ne.jp/~mkty/	62 · **63**
PACHI PACHI	天野かおる	http://home3.highway.ne.jp/pachi-2/	90 · **91**
Page5	FIFTH	http://www01.u-page.so-net.ne.jp/db3/fifth/	48 · **49**
PANDORABOX	ごんすけ	http://www.win.ne.jp/~gonsuke	70
PARAIBA TOURMARINE	石田真由美	http://member.nifty.ne.jp/pbm/	64
Pastel Image	SPHERE	http://www.tt.rim.or.jp/~sphere/	58
Pastime	ちょも	http://www7.peanet.ne.jp/~cho/	82
Peeping P!	ごとP	http://www.mirai.ne.jp/~gotop/	122
Persian Pudding	天野 忍	http://home2.highway.or.jp/naono/	64
Phantasmagoria	紗那	http://www2s.biglobe.ne.jp/~SANA/	42 · **43**
PHANTASYSTEM	かごめ	http://home.highway.or.jp/yosshi/main/kagome.htm	14

Phase Green	川原由唯	http://www.ricoh.co.jp/src/people/fukuhara/	40
Pi's HomePage	ぴー	http://netv5.netvision.co.jp/~pi/	18
piano:forte a.s.c.	氷澄水	http://www.din.or.jp/~hisumi/	28
PING's WING	ぴんぐ	http://web1.tinet-i.ne.jp/user/chinke/	92・**93**
PINKGUN	あずなしあ	http://www4.big.or.jp/~azuna/pinkgun/	54
pit_inn	ぴぃ♪	http://www.lares.dti.ne.jp/~dfa/pit/	84
plumroom	plum	http://www2s.biglobe.ne.jp/~plumroom/	84
POINT BLANK	山椒 奈々実	http://plaza15.mbn.or.jp/~sansyou/	24
PoisonArts 致命傷/FATALISM WORKS	弥舞秀人	http://www.ipc-tokai.or.jp/~hyde/	114
PON'S-BRAND	大乃国ぽん	http://www.alles.or.jp/pub/nezupon/	64
ponk The Funkey Bunney	梅田"ACTY"大丸	http://www.tk.airnet.ne.jp/acty/ponk/	114
POT☆CRAFT	ぽてきち	http://www.ask.or.jp/~mao/	72
POWWOW	樹華	http://www.enjoy.ne.jp/~jyk/	84
PRECIOUS	ふうきりえ	http://www.bekkoame.ne.jp/~imashiro/	28
Propaganda	九石はくね	http://www2.wbs.ne.jp/~system9/	82
PSY [sai] in DiSiTaL WoRKs	PSY [sai]	http://www.os.rim.or.jp/~psy/	44
PURPLE CASTLE -in MATSUKAZE Pub.-	麵次	http://w32.mtci.ne.jp/~rito/	64
quattro Hobbs Room	くわとろ	http://www02.so-net.ne.jp/~quattro/	44
QUEERNESS	Nayu	http://www.alles.or.jp/~nayu/	10
R Project	瑠堂れおな	http://www02.u-page.so-net.ne.jp/xb3/ryu-sei/	96
R's ROOM「落描きのススメ」	MRR	http://plaza29.mbn.or.jp/~mururu/index.htm	10
Rakia CG Garden	ラキア	http://www.asahi-net.or.jp/~ty5a-kmr/	20
ramble's home page	ramble	http://home4.highway.ne.jp/ramble/index.htm	10
REPORT	日高司	http://www.os.rim.or.jp/~report/	44
Retu's Home Page Cocktail	比十之 烈	http://www.ggf.ilc.or.jp/user0/retu/	84
Rhapsody	K.Shiina	http://www2q.biglobe.ne.jp/~SHIINA/	36
Ring Ring Mix!	りんだ	http://www.din.or.jp/~rinda/	44・**45**
RINN's Page	Woody-RINN	http://www.din.or.jp/~rinn/	12・**13**
Ritzve's Dreamscape	りつべ	http://www.246.ne.jp/~ritzve/	22・**23**
Rock Climbing	岩舘こう	http://www.ceres.dti.ne.jp/~iwadate/	58・**59**
Rosanna's Homepage	Rosanna	http://www.asahi-net.or.jp/~mt3t-njr/	10
rosette	あいざわひろし	http://www02.so-net.ne.jp/~highrisk/proom/aizawatop.html	58
Ruby's Page	Ruby	http://village.infoweb.ne.jp/~ruby/	70
Ryo's Collection	Ryo	http://home.interlink.or.jp/~ryo1/	90
SAFIRE	SAFIRE	http://www.bekkoame.ne.jp/i/safire/	112
SALTSHAKER	塩原信一	http://www.alles.or.jp/~s2r/	46・**47**
SANDWORKS	砂	http://www.campus.ne.jp/~suna/	102
SANSHIRO AKIYAMA'S Web page	346	http://www.din.or.jp/~sanshiro/	**36**
Saparak	さぱら	http://www.yo.rim.or.jp/~sapara/	44
SATOXのホ~ムペ~ジ	SATOX	http://www2s.biglobe.ne.jp/~satox/	44
SCHIZOPHOTIC	佐倉修造	http://plaza29.mbn.or.jp/~SAKURA/	66

System町娘	つるぎゆきの	http://www.phoenix-c.or.jp/~yukino/	102
T.O.H.P.	津霧みん	http://www.din.or.jp/~tugiri/	**42**
T.R.Y-ZONE	羅陰知英	http://www.sun-inet.or.jp/~uiy00810/	84
T'ienLung (ティエンルン)	さとをみどり、どりんち	http://plaza29.mbn.or.jp/~tienlung/	92
Tajima Show Room	たじまはるみ	http://www.cyborg.ne.jp/~kana/	122
TAKA's on Web.	TAKAHiCo	http://www.netlaputa.ne.jp/~takas/	84
Takam 作品集	Takam	http://www.ask.or.jp/~takam/	12
TAKAMICHI FACTORY	たかみち	http://www.amy.hi-ho.ne.jp/takamichi/	16
Takapen & NijimasuX's HomePage	Takapen	http://w33.mtci.ne.jp/~takapen/	120
TAP WORLD	たっぷ	http://www.asahi-net.or.jp/~je6t-fjt/tap/	98
TARAKO FACTORY	たくま朋正	http://www1.pos.to/~takuma/	46
TARGO's WorkShop	TARGO	http://www.tt.rim.or.jp/~tagro/	58
TATSUKobo	たつのすけ	http://www.vector.co.jp/authors/VA012144/	110
TEA·ROOM KO-CHA's Web Page	こ〜ちゃ	http://www.din.or.jp/~ko-cha/	46
TECHNO LIFE	松野唯	http://www02.so-net.ne.jp/~y_co/	58
The Critical Point Of Desire	金米糖	http://www.bekkoame.ne.jp/ha/konpeito/	80
The Little World of jaja	じゃじゃ	http://www.yk.rim.or.jp/~jaja/	84
time.	きすき ゆずる	http://www.geocities.co.jp/Playtown/4029/	50
TK150's Miscellaneous Area	TK150	http://www.asahi-net.or.jp/~SD3T-KTU/	98
To. Digital ARTs.	To.	http://www.vc-net.or.jp/~tosiyuki/	88
Tokyoサーキット	delica	http://anya.org/t_circuit/	108
TOMORAの生けどりInternet	智羅	http://www.big.or.jp/~tomoki/	24
Toshi-B's image Factry	トシB	http://www.atnet.ne.jp/~toshit/	16
ToSINo Gallery	ToSINo	http://www3.osk.3web.ne.jp/~tosino/	98
TOTEN's ROOM	トーテンコプ	http://www.bekkoame.ne.jp/~toten/	112
TOYPOP IN WAVE	shura	http://www.din.or.jp/~shura/	96
TRAIN OF THOUGHT	3PAY HIRO	http://www.sun-inet.or.jp/~sanpei/	54
Trush&Groria's Home Page	Trush	http://www.infonia.ne.jp/~fix/	110
Twilight Aquarium	川羽らみき	http://www2.justnet.ne.jp/~ramiki/twilight/	122
Twinkle Town	覇王絹丸	http://www.future.ne.jp/kinumaru/	62
UN⇔NET	雲竹彩ーN	http://www2.117.ne.jp/~unnet/	84
Underground Sample Files	孤児郎	http://plaza19.mbn.or.jp/~kojie/USF/	112・116
UniUni's HomePage	うにうに	http://www.venus.dti.ne.jp/~uniuni/	46
UNSTABLEST DESIGNxxx	カイ	http://www.geocities.co.jp/SiliconValley-PaloAlto/1315/kai_index.html	52
Untitled-1	As'257G	http://www.yk.rim.or.jp/~as1130/	84
UpperD3 On Line	kimtoy	http://www.mars.dti.ne.jp/~dggc/	122
VALKYRIE ILLUSION	士貴貴ゆう	http://www.vc-net.or.jp/~vishiki/	46
VANGUARD FLIGHT	押野 卓	http://stu.nit.ac.jp/~e977006/	98
Vertigo High	ほづみないぎ	http://www.venus.dti.ne.jp/~nhodzmi/	52・**53**
Vortex of Arts.	水谷わひょみ	http://www.nue.net/~wahyomi/	122
wakachan's HomePage	和佳-chan	http://wakachan.kaynet.or.jp	48

SchwarzeKirschen	21世紀計画社	http://www.yo.rim.or.jp/~takeori/sk/	72
Scrap machines	大王	http://www7.peanet.ne.jp/~dai_nn/	48
SCRATCH	たかぼ	http://hiroba.net/scratch/	16
Seraphic	砂月雪人	http://csefs01.ce.nihon-u.ac.jp/~u086077/	66
Seto Home Studio	Seto	http://member.nifty.ne.jp/seto/	12
SEVEN FLAVOR SPICE	しゅんにい	http://www2.tky.3web.ne.jp/~syunichi/	46
SHIROのお絵書き工房	SHIRO	http://www.bekkoame.ne.jp/~y.shiro/	12
SHO ROOM	松本彰	http://www.asahi-net.or.jp/~rd8t-mtmt/	60
SHOU's GALLERY	sho	http://www.alphatec.or.jp/~shouta/	12
Sion's Web Page	紫音	http://www.alles.or.jp/~karin/	66
Site Take-W	武W	http://www.alles.or.jp/~takew/	86
SITE-COCO	コイデココロ	http://www.sanynet.ne.jp/~n-n-e/	84
SKILL SEEKERS	TAKAHIRO	http://www.kulawanka.ne.jp/~takahiro/skill/index.html	120
SMASHING PUMPKIN'S HOMEPAGE	さむな	http://www.fuji.ne.jp/~sumner/	48
SnowHouse	ユキ	http://member.nifty.ne.jp/Yu-Ki/	92
Soft Page～やわらかいものすきですか？～	ミズクラゲ	http://www.alles.or.jp/~kurage/	20
SOG club	氷優きゃあ	http://members.aol.com/rinrin201/hp/Untitled1.html	28
sonson's Cafe	そんそん	http://www.win.ne.jp/~sonson/	50
SPR NEWS	さとP	http://www.mars.dti.ne.jp/~sato-p/	84
Squid Trap	まっつあん	http://www.osk.3web.ne.jp/~ma2an/	42
still I believin	RYUNE	http://www.asahi-net.or.jp/~SY8K-HTNK/	80
StOneBOX(すと～んぼっくす)	野沢醒矢	http://www.pluto.dti.ne.jp/~seiya-n/	50
Studio Cycles of Life	NMR	http://www.sm.rim.or.jp/~nmr/	122
STUDIO JES HOME PAGE	JES	http://village.infoweb.ne.jp/~jes/	72
STUDIO JIP!	JINPINO	http://home.interlink.or.jp/~tkumagai/jinpino.html	72
Studio LOGIC	PANTO MAIMU	http://www.asahi-net.or.jp/~gu7m-myk/	28
STUDIO M	久川 晶	http://stein.qse.tohoku.ac.jp/~hekiru/	72
Studio Myu.	山ちゃん	http://www.kt.rim.or.jp/~myu2/	62
studio necohouse	ねこのにゃお	http://www.fitweb.or.jp/~nyao/	122・**123**
STUDIO S.D.T.	結城辰也	http://www.sic.shibaura-it.ac.jp/~l96073/	102
STUDIO SUNNY SPOT	ひなた·ゆきね	http://www02.so-net.ne.jp/~yukine/	98
Studio Trash Official Homepage	Studio Trash	http://plaza17.mbn.or.jp/~STrash/	70
STUDIO VANGUARD	安童あづ美	http://plaza13.mbn.or.jp/~studiovanguard/	112・116
STUDIO VIEWPORT	水瀬晶史	http://www.asahi-net.or.jp/~ub3k-twr/	92
STUDIO ZERO	JACK!zero	http://www.studio-zero.com/	48
STUDIOふあん	来鈍	http://home3.highway.ne.jp/~huan/	90
STUDIOぶーびーとらっぷ	STUDIOぶーびーとらっぷ	http://member.nifty.ne.jp/TOKUMEIKA/	70
Studioぽんぐろまりっと	ぽんぐろ	http://netpassport-wc.netpassport.or.jp/~wsonehar/	106
STUDIO亜人類	小室恵佑	http://www.bekkoame.ne.jp/~k_komuro/ajinrui.html	76・**77**
SUZUKEN	鈴木 健	http://member.nifty.ne.jp/suzuki-ken/	28
SYNTHESIS	TASH	http://w3.mtci.ne.jp/~tash/	12

Water Colors	ayako	http://www.dokidoki.ne.jp/home2/ayako/	88
Web Potahouse	SAS_P	http://plaza3.mbn.or.jp/~pota/	10
Welcome to the Rose Garden	樫夢	http://www.akira.ne.jp/~kasimu/へ移転予定	40 · **41**
WHITE CREATION	増田幸紀	http://www2s.biglobe.ne.jp/~ym_page/art/	58
without sun	あいね	http://www.age.ne.jp/x/tadashi/	14
Wonder-Ranch	いとけい	http://www.sainet.or.jp/~itokei/	52
WOODY HILL	狼犬太	http://home.att.ne.jp/gold/kenta/	102 · **103**
XANTHUS	XANTHUS	http://www.din.or.jp/~hikayu/	14 · **15**
XANTHUS & PAPILLON	XANTHUS & PAPILLON	http://www.top.or.jp/~xanthus/	14
xeronion's work	xeronion	http://deniam.com/users/xeronion/	66 · **67**
Y's Graphic factory	ワイズ	http://plaza6.mbn.or.jp/~YGF/	20
Yagiyama Publishing	いちせ	http://www2e.biglobe.ne.jp/~ichise/	84
YAMA's HomePage	YAMA	http://www.crt.or.jp/public/user/~yama/	16 · **17**
Yeemar's Home Page	Yeemar	http://www.asahi-net.or.jp/~QM4H-IIM/index.htm	12
YouMa in Coterie	ゆうまじろう	http://www.lares.dti.ne.jp/~youma/coterie/	46
YOUMEI's home page	Youmei	http://aumy.biwako.shiga-u.ac.jp/std/s96762hw/youmei.htm	14
yugori page	yugori	http://www.asahi-net.or.jp/~MN2Y-TBT/	68 · **69**
Yuris cafe	椎名悠理	http://www.fsinet.or.jp/~yuris_c/	86
YUSER'S倶楽部	TM·ゆ〜ず	http://www2.justnet.ne.jp/~h_yuse/	94
Z·X·カンパニー	DON	http://www2s.biglobe.ne.jp/~don-a/	88
ZANY	倉本なつね	http://www.geocities.co.jp/Playtown-Denei/8325/	24
Zero Second	安森 然	http://plaza19.mbn.or.jp/~zen_yasumori/	20
ZX-RR CG PAGE	Rai	http://www.intacc.ne.jp/~gisho/	12
あい·ぼうるアニメCGホームページ	あい·ぼうる	http://www.kt.rim.or.jp/~eyeball/	70 · **71**
あおいNET	MIKAMIKA	http://www3.justnet.ne.jp/~mikas/	86
あおきなおウェブページ	あおきなお	http://www2s.biglobe.ne.jp/~AOKINAO/	54
あしゅらのおもちゃ箱·はいぱ〜!	雅あしゅら	http://www02.so-net.ne.jp/~asyura/	62
あとりえRUURY	RUURY	http://www.bekkoame.ne.jp/~ruury/	70
ありすの喫茶店	秋月あかね	http://www.din.or.jp/~akiduki/	22
あんも·ないとの部屋	あんも·ないと	http://www.asahi-net.or.jp/~qn9m-ysd/ammonite/	26
うさうさFのホームページ	うさうさF	http://www.bekkoame.ne.jp/i/fujisaki/	26
うさぎ屋本舗	妹尾ゆふ子	http://member.nifty.ne.jp/usagiya/	92
うるさい小娘	とりのすけ	http://www.raidway.ne.jp/~iharay/	120
えいくんち	えいくん	http://www.capricorn.cse.kyutech.ac.jp/~ei/UC/	62
えりくしる	紅竜鷹羽	http://home3.highway.ne.jp/nisa/	20
おえかきほーむぺーじ	かとうよしとも	http://village.infoweb.ne.jp/~katoh/	70
おーだーメイド(closed)	SOARER	http://www.angel.ne.jp/~rinse/	102
オーバーオール友の会	みしきさかな	http://www.sfc.keio.ac.jp/~t96639yn/mishiki/overalls.html	102
おやじん家	おやじ	http://www.dango.ne.jp/~gfa01575/	30
がとーのStudioBEYOND	がとー	http://super.win.ne.jp/~gato/	16
かにさんネット	かにさんネット	http://www.catnet.or.jp/kanisan/net/	118

カフェテリアWATERMELON	kosuge	http://plaza24.mbn.or.jp/~watermelon/	112・116
がんちゃんの3DCGのページ	岩尾眞吾	http://www.aikis.or.jp/~s-iwao/	98
きいちご魔法店	AOI☆	http://www.vector.co.jp/authors/VA004239/	110
きつね友の会公民館	delica	http://anya.org/fox/	108
きゃろっと美少女パラダイス	てれうさ	http://home6.highway.ne.jp/teleusa/	102
キラキラ★ヒカル!	さっぽろモモコ	http://www.cc.rim.or.jp/~momoko/	16
グラタン帝国	真黒さしみ	http://www-user.interq.or.jp/~sashimi/index.html	34
くろうさ工房	oku	http://www.vector.co.jp/authors/VA008796/	10
げーむどーじんりんく	武折貴子	http://www.ohkini.net/~takeori/gdl/	120
ケン太+樹のCG「野望のページ」	ケン太&樹	http://www.scarecrow.co.jp/~yabou/	58
ことばの国	COTOBA	http://www.baba-t.com/~cotoba/	62
ごりぽん協会ホームページ	ごりぽん協会	http://www.ask.or.jp/~seishiro/goripon.htm	16
サディスティック・アルケミィ~嗜虐的錬金術~	青文鳥	http://www.din.or.jp/~aobun/gate0.html	114
しのさん美術館	しの	http://www.ddt.or.jp/~shino/	74
しのだよしたかのホームページ	しのだよしたか	http://kenoh.hits.ad.jp/~kenichi/	70
しめじのまぜごはん	秋野しめじ	http://plaza8.mbn.or.jp/~gosui/	92
しゃんふぁ・なにむ	水明&瑞浪かさね	http://www.246.ne.jp/~miz/	22
しゅう・とくとみの忍者貴族	しゅう・とくとみ	http://www.asahi-net.or.jp/~iu7k-eczn/	16
スタジオ DELTA	SPRIGGAN	http://www.os.xaxon.ne.jp/~spriggan/	86
スタジオ インパクト	スタジオ インパクト	http://ha2.seikyou.ne.jp/home/harumaki/	110
すてすて~Stay Steady~	とむそおや	http://www.at-m.or.jp/~visnu/	36・**37**
するめや	Oh-Shall & TEN	http://www.asahi-net.or.jp/~FP1Y-OONS/	10
たかしろ亭	たかしろそうま	http://plaza12.mbn.or.jp/~takasiro/	16
たかはし倶楽部	たかはし	http://plaza18.mbn.or.jp/~s_taka/	68
ちぬちぬ少女の王国	SNAKE-PIT	http://www.bekkoame.ne.jp/~snake-pit/	116
ちゃ~むなねっと	You	http://w3ma.kcom.ne.jp/~k-ogawa/charm/	14
ディラパル　ブラウザ	代由	http://member.nifty.ne.jp/tsunoda/	64
どきどきくじらんど	火延 真	http://www.cc.rim.or.jp/~hinobe/	62
ときめき__BOX	狼太郎	http://www02.so-net.ne.jp/~wolf/	60・**61**
とった~くんの部屋	熊木十志和	http://www.bekkoame.ne.jp/i/kumagon/	90
とまとやホームページ	冷やしたとまと	http://www.aaa-int.or.jp/tomato/	64
ナミフクDM	ひなた☆すう	http://home.interlink.or.jp/~hinata/namifuku/	118
ねことこWorks	ねこねこ&とことこ	http://nekoneko.com/nekotokoworks/	56・**57**
ねこみみ振興会	いんくぽっと	http://www.cc.rim.or.jp/~inkpot/	106
ねこわんこホームページ	ねこねこ	http://www.asahi-net.or.jp/~jp8k-iwsk/	26
のにさんのホームページだ!	のにさん	http://csefs01.ce.nihon-u.ac.jp/~u086119/	18
のりのホームページ	のり	http://plaza15.mbn.or.jp/~norino/	68
はあとぎゃらりー	今井一成	http://members.aol.com/SASAMIC43/	66
パイナップル畑	桜餅 智	http://www.alles.or.jp/~tomos/	66
ばな~倶楽部	Woody-RINN	http://www.din.or.jp/~rinn/bnnr/	118
ハニー ソノちゃま らぶらぶっこ倶楽部	三上 ソノ	http://www.alles.or.jp/~honey/	98

はにわ南蛮	花形水琴	http://home3.highway.ne.jp/h-mikoto/	26・**27**
ぱやしのページ	ぱやし	http://www.asahi-net.or.jp/~fv3n-wkby/top.html	118
はるのつぼみ	つぼみしゅん	http://www.big.or.jp/~t-shun/index.html	16
ぱんちゅ万歳	h_k	http://www.asahi-net.or.jp/~tr6h-kjur/	36
はんな屋	はりけんはんな	http://www.bekkoame.ne.jp/i/hannaya/	112
ひじりまあみのお部屋	ひじりまあみ	http://home2.highway.ne.jp/maami/	68
ひまわりカンパニー	七瀬いーうぃ	http://www.asahi-net.or.jp/~JL6T-NKGW/	22
ひろいずむ	大嶋ひなと	http://www2u.biglobe.ne.jp/~hinato/	24
ふつーのほーむぺーじ	こうひい	http://www.cyberoz.net/city/kohi/index.html	62
ふらふら通り	ばいばい	http://www.ddt.or.jp/~baibai/	104
プリンシュー	須田さぎり	http://www.bekkoame.ne.jp/ha/sagiri/	22
へっぽこギャラリー	りとるぐれい	http://home3.highway.ne.jp/banana/	90
ぺぱーMintグリーンホームページ	安木美恵子	http://village.infoweb.ne.jp/~fwhe8793/	20
ほぉりぃ☆べるのお絵カキの～と	ほぉりい☆べる	http://www.bekkoame.ne.jp/~beldandy/HolyBell/	116
ほかほか書店	ほかほか書店	http://www.amy.hi-ho.ne.jp/taro-chiaki/	74
ぽこぽこダイナマイツ	Chororphis	http://plaza26.mbn.or.jp/~chorop/	8
ほたる狩り	恋純☆ほたる	http://w3.mtci.ne.jp/~hotaru-k/	26
まぐまぐパラダイス	まぐまぐソフト	http://magmag.kaynet.or.jp/	118
まくらの部屋	夢里まくら	http://plaza6.mbn.or.jp/~makura/	106
まさあきのCGギャラリー	まさあき	http://www.ap.kyushu-u.ac.jp/appphy/member/ooishi/index-j.html	18
まじかるすてーしょん	LESIA	http://www.dd.iij4u.or.jp/~lesia/index.html	26
まじかるメロディ	Melody-Yoshi	http://www.alles.or.jp/~melody/	78
まどかぱ～く!	膜ちゃん	http://www.cute.or.jp/~makuchan/	86
まどかぱ～く! 別館	膜ちゃん	http://www.ra.sakura.ne.jp/~pelmo/madoka/	86
ミノンのお部屋	NON	http://www.geocities.co.jp/Playtown/5676/	52
みぽ゛ずシステム【MIF】	黒猫	http://www.geocities.co.jp/Playtown-Denei/8930/Homepage.html	108
めいどいんぢゃぱん	眉毛	http://www.asahi-net.or.jp/~nu7h-ootw/	100・**101**
めいぷる　りーふれっと～Kaede's CG Gallery	かえで☆	http://www2s.biglobe.ne.jp/~kaede_p/	36
めいへう゛ん	ぼいどめいん	http://home.intercity.or.jp/users/onishi/	18
めがねがね	はいぼく	http://www.bekkoame.ne.jp/ro/haiboku/	100
もちもち本舗	EXP	http://miya.or.jp/~mochiex/	52
よこむきもぐら	ハムくまズ	http://www.angel.ne.jp/~nochina/	62
らpcon	くま坂らま男	http://www.cosmos.co.jp/~ramao/	116
らいとっ!	のん	http://plaza8.mbn.or.jp/~nonn/	18
ランジェリンク	398	http://www2n.biglobe.ne.jp/~syoujyor/	120
ルートS.S.のハッピーカムカム	Route S.S.	http://cf.vow.co.jp/rss/	10
るりんく	398	http://sa.sakura.ne.jp/~syoujyor/	120
るるーず推進委員会	デンタゴンるる	http://w3ma.kcom.ne.jp/~hi6/	68
レイバー少女産業振興協会	Ka2DA	http://www.mars.dti.ne.jp/~ka2da/labor_girl/index_lg.html	88
ろりろりメルトダウン	左向き矢印	http://home2.highway.ne.jp/socket5/	**44**
絢爛漫画遊戯館宮野式	宮野あみか	http://member.nifty.ne.jp/miyanon/	66

綾丸堂本店	龍王綾丸	http://w3.mtci.ne.jp/~ayamaru/	26
暗黒太陽通信	藤川純一	http://member.nifty.ne.jp/devilkitten/toppage.html	26
烏賊川通信社	perpasia	http://www.bekkoame.ne.jp/~perpasia/index.html	90
音箱	りどる	http://www.bekkoame.ne.jp/ro/ridoru/	92
化け物アイランド	メカうほほ1号	http://www.annie.ne.jp/~meka/	20
仮想的小葉～Virtual Leaflet～	いわは	http://www.cyborg.ne.jp/~iwaha/	76
霞草館	山火 霞	http://www.aianet.ne.jp/~yamabi/	90
我楽多市	相生葵	http://www.din.or.jp/~aioi-aoi/	74
魁!!アニメ塾	MASA	http://www.fsinet.or.jp/~stms/	78
海カラス共和国	ゲルニカ	http://www.bekkoame.ne.jp/~guernica/	96
海星★梨の落書きホームページ	海星★梨	http://member.nifty.ne.jp/hitode/	20・**21**
学園はにもくお	学園はにもくお	http://club.infopepper.or.jp/~hatahira/	88
楽天主義国	藍雨絵緒	http://www.din.or.jp/~eoaiu/	30
喫茶「ひだまり」	ひなた☆すう	http://home.interlink.or.jp/~hinata/	18
喫茶赤べこ屋	kami	http://www.geocities.co.jp/Playtown-Denei/2156/	74
橋本組出張課託児所	兼越ゆうじ	http://hot.netizen.or.jp/~akari/	66
狂乱麗舞	光明公	http://member.nifty.ne.jp/morii/	66
暁宙館★☆	暁宙	http://www.aitech.ac.jp/wing/~meaken/gyouchu/welcome/	92
近道。	かせん	http://plaza26.mbn.or.jp/~chikamichi/	16
銀龍牙'sほうむぺえじ	銀龍牙	http://www.alles.or.jp/~gin/INDEX.HTM	66
建国計画	騎羅	http://www.top.or.jp/~kiran/kenkoku	92
犬飼神社	真田丸マサオ	http://plaza28.mbn.or.jp/~inukai/jinjya/	66
元気薬	HIRO PON	http://www02.so-net.ne.jp/~hiropon/	8
幻影図書館	ミラージュ	http://plaza28.mbn.or.jp/~miragenovels/	116
幻想の星空	翔空聖流	http://www.ask.or.jp/~mtint/fsn/	118・**119**
幻想王国	Funny.	http://www.alles.or.jp/~funny/	96
工房猫の手	あらきよう	http://www.asahi-net.or.jp/~nx1m-stu/	14
溝川淀美の部屋	溝川淀美	http://www.bekkoame.ne.jp/ro/yodomi/	60
香月柚奈ほーむぺーじ·こざるのゆなちゃん	香月柚奈	http://www.bekkoame.ne.jp/~keen/yuna/	22
黒猫館	篠崎広里	http://plaza4.mbn.or.jp/~kuronekokan/	66
此処あ夢工房	夢工房	http://member.nifty.ne.jp/YUME/	92
魂の兄弟たち	よしいさん	http://member.nifty.ne.jp/yyoshida/	68
佐倉千歳堂	佐倉千歳	http://www.246.ne.jp/~c-sakura/	22
佐藤和芳アートギャラリー	佐藤和芳	http://www.aland.to/~e-ksato/	22
三つ編み萌え	紫 らない	http://www.alles.or.jp/~walkure/	66
山野草	雪原 露	http://www1.akira.ne.jp/~shoma/	114
持続力-C	持続力-C	http://www.seaple.icc.ne.jp/~dai/hp/	**38**
時雨堂	水月楓	http://www1.odn.ne.jp/kaede/	44
自然を大切に	syo	http://www.ceres.dti.ne.jp/~syo/	70
秋月書店	秋月つき	http://home.interlink.or.jp/~akituki/	60
緒計画	VIN緒	http://www02.so-net.ne.jp/~vincho/	98

小手川ゆあの極楽刑務所	小手川ゆあ	http://www.bekkoame.ne.jp/ro/gj13041/	34
小田修紀CG館	小田修紀	http://village.infoweb.ne.jp/~fwgi0240/	22
少年グラタン	茶柱たつや	http://www.click.or.jp/~cha/	30
少年的電影箱	瑠堂れおな	http://www.geocities.co.jp/Playtown/8756/	40
上海楼	佐久嶋火急	http://www.alpha-net.or.jp/users/sakujima/	100
新あおしま製菓	青島ういろう	http://web.kyoto-inet.or.jp/people/aoshima/	24
水精少女譚	ななみ	http://www-user.interq.or.jp/~nan/	18
世界猫硝子館	遊魔海里	http://www.meix-net.or.jp/~noah/	84・**85**
星影美術館	濱口よしたか	http://member.nifty.ne.jp/yositaka/	96
仙道ますみのシタゴコロ	仙道ますみ	http://www.dmp.co.jp/masumic/	30
閃骸境	ありさわとよみつ	http://village.infoweb.ne.jp/~fwie0190/	14
全国庁 にこりん情報局	cue	http://www.ainet.or.jp/~cue/	8
村上水軍の館	村上水軍	http://www.alles.or.jp/~msuigun/	100
第三新東京星　天使居住区	嵩石恭巳	http://member.nifty.ne.jp/yasumi/	64
第七南瓜部隊	KISKE	http://www.age.ne.jp/x/kiske/	116
地雷屋	MINE	http://www.ceres.dti.ne.jp/~mine-/	114・**115**
朝	高木朝成	http://www.angel.ne.jp/~wawon/	74
泥酔桜国	まると!	http://www.luvnet.com/peoples/maruto/	**108**・**109**
天空の杜	智樹	http://club.pep.ne.jp/~apt.hosono/index.html	96
天使の涙	くたぽでぃ	http://www.bekkoame.ne.jp/i/fl2106/	90
天然色	井上海松	http://www.bekkoame.ne.jp/~iiyoshi/	20
天体少女図鑑	ななみ	http://nanami.px.to/	18
電気ポット	攻牙沙	http://www.bekkoame.ne.jp/i/fl2179/	90
東風	ばいと屋某	http://www.ask.or.jp/~byte/	18
藤岡建設	藤岡建機	http://www.ceres.dti.ne.jp/~omecha/	62
特盛EXPRESS	吉野家うっしー	http://member.nifty.ne.jp/yoshinoya/	20
那由他画堂	那由他	http://ha5.seikyou.ne.jp/home/nayuta/nayuta_gado/	30
猫族集会	KFT	http://www7.big.or.jp/~kft/NEKO/index.html	8
猫秘密情報結社	岩崎れえこ	http://www.246.ne.jp/~reiko-h/	20
脳みそうにうに劇場	太田まさよし	http://w3.mtci.ne.jp/~myosi/	**104**
爆熱FAM_Activate()	FAM柳瀬	http://www.age.ne.jp/x/fam/	8
爆裂健 Home Page I	爆裂健	http://www1.odn.ne.jp/~aaa33460/	110
爆裂健 Home Page II	爆裂健	http://www.vector.co.jp/authors/VA006860/	110
麦酒店舗	ぱぴよん	http://www.din.or.jp/~beershop/	18
八尋殿	朝倉純壱	http://plaza25.mbn.or.jp/~junchiro/	24
秘書室インターネット版	秘書	http://www.bekkoame.ne.jp/ha/masako/	26
秘密倶楽部ほぉ～むぺぇ～じ	Syun＆星河☆苗	http://plaza7.mbn.or.jp/~secret_club/	110
秘密結社Neo-Sea-Hoese	魔導師うっちー	http://www.bekkoame.ne.jp/~utty0	90
緋村屋	kami	http://www.awave.or.jp/home/naga1615/himuraya/	74・**75**
非万能電化研究所	miho	http://village.infoweb.ne.jp/~fwix7358/	10
備後萬屋	さいとうつかさ	http://www.catnet.ne.jp/yorozuya/	30

備前屋	ひの	http://s3.hopemoon.com/~elx001/	68
微笑み人形の館	ひの	http://www4.justnet.ne.jp/~dolls/	18
美紗魅	猫艦¢	http://www.cc.rim.or.jp/~nekokan/	106
美少女動画同人ソフト C.A.T.	C.A.T.	http://karakuri.com/	110
氷猫は冬眠中	氷猫ICE	http://plaza11.mbn.or.jp/~icecat/	106
瓢箪本舗	佐伯涼子	http://village.infoweb.ne.jp/~saeki/	88
風の森工房	マサ	http://user.shikoku.ne.jp/hmx12mlt/	94・96・**121**
風雀のHP	風雀	http://www.asahi-net.or.jp/~SJ8T-YSMT/	64
碧の館	海女原磨莉	http://www2.justnet.ne.jp/~rskj/	80・**81**
穂高んちの作品掲示部屋	穂高ひとみ	http://web.kyoto-inet.or.jp/people/htaka/	74
豊楽煩悩館	豊楽	http://plaza12.mbn.or.jp/~horaku/	100
夢幻飛行	瑞枝 悠	http://plaza3.mbn.or.jp/~mugen_hikoh/	56
夢色鉛筆	まさる	http://www.asaka.ne.jp/~masaru/	62
夢戦士ドリームソルジャーZ	System 3(ag)	http://cgi.din.or.jp/~ta2chi4/dream/	12
娘細工	ゆきの	http://www.osk.3web.ne.jp/~yukino/(現在更新停止中)	20
迷宮美術館	夕凪雄那	http://www.asahi-net.or.jp/~cz6t-sm/	26
滅砕CGROOM	JMS	http://www.asahi-net.or.jp/~zj3y-situ/cgroom.html	90
妄想画廊	高橋ダム	http://www02.so-net.ne.jp/~dam/	114
木曽屋本店	木曽秀平	http://www.blk.mmtr.or.jp/~shuuhei/	26
目んたまTシャツ王国	NMR	http://www.sm.rim.or.jp/~nmr/GAME/magic/	122
目薬専門店ねぼけ堂	かわち丸	http://fame.calen.ne.jp/~kawachi	16
友遊広場	ANNANDULE Project	http://www.na.rim.or.jp/~fre/anan/	110
夕凪堂本舗	東雲あずみ	http://ha1.seikyou.ne.jp/home/yuunagi/	26
蘭香学園高等部	人造人間s707号	http://w33.mtci.ne.jp/~s707/	114
裏・百鬼夜行	ましたか	http://www.bekkoame.ne.jp/~mashitaka/	102
零の光景	みかげ	http://www.bekkoame.ne.jp/ha/mikage/	68
惑星たごや	竜にょ	http://member.nifty.ne.jp/riy/	96
嗚呼、我等加藤隼戦斗隊	加藤	http://www.mars.dti.ne.jp/~naitou/	**106**・**107**
嗚呼！極限劇場	高木裕司	http://home.intercity.or.jp/users/accell/	24・25
篝火工房	水谷ヨージ	http://home.att.ne.jp/red/miz/	64

後記

在寫完本書之後，筆者決定在刊登標識(banner)或CG作品時，務必取得原作者的同意，並且請原作者說明製作標識(banner)或CG作品的概念。雖然這是理所當然的事，但過去卻一直受到出版業界所忽視。

筆者使用電子郵件，請求原作者同意我刊登他們的作品，我約發出一千二百封的電子郵件，而收到九百封的回函。大部分的原作者都接受筆者的詢問，也同意讓我刊登他們的作品。可是，在我閱讀了回函時，發現不少原作者對商業雜誌的印象極為惡劣，這是因為商業雜誌不重視上述程序的緣故。

為了消除他們對商業雜誌的不良觀感，筆者無微不至地回應他們的詢問與要求，同時也隨時接受他們的意見。但因為這個緣故，使我們的編輯方針一改再改。另外，由於我的失誤與誤判，引發了一位原作者的不快。在這裡，筆者深深地表達萬分的歉意。如果我有什麼不懂的地方，請各位不要客氣，直接指出來。雖然我盡心盡力地想要做好本書，不料卻招致這樣的結果，這是我需要確實反省的地方。

我已經開始將標識(banner)當作一件作品來看待，深深地覺得標識(banner)的背後，存在著原作者這個事實。並且，體認到理解將近一千位原作者之意思的重要。

就這樣，我們掌握了本書的編輯方針，同時也盡可能地做到精益求精的地步。可是，有些地方還是不盡滿意。我本來有重作本書的想法，但因為必須按照預定的期限出版，在某種程度之內，只好與現實妥協。

我想，同意讓作品刊登在本書上的原作者，在實際看了本書之後，應該會有不滿意的地方，尚祈各位不吝賜教，以供我們再版的參考。

雖然有過妥協，但我認為本書的出版也算是一項成果。至少我已經按照當初的想法，將本書呈現在讀者的面前。

直至截稿日期為止，我們刊登的標識(banner)大約有一千幅左右，大概不到全體的三分之一，我們希望透過這次的經驗，能夠把第二版做得更好。

對於各位CG作家的指教、協助與鼓勵，在此深表謝忱。

最後，我要感謝古拉費克出版社給我這個機會，讓我的想法能夠成為書本問世。我還要感激奧田小姐、翻譯Kristin小姐、野澤小姐，以及在精神上一直支持我的Mywa小姐，還有許多好朋友們。

Behind the Banners are People

In the making of this book, I decided to explain the concept behind the book to the artists concerned and obtain their permission for each banner and CG artwork to be published. This may seem obvious, but I state this here because it is something that is often neglected in the publishing world.

I used the Internet to obtain the artists' permission. I sent about 1,200 E-mail letters and received almost 900 responses in reply. Most of these were from people who supported the book and consented to me publishing their artwork. However, reading through their replies, the impression I got was that many seemed to have a low opinion of people working at commercial magazine companies. Perhaps this shows that publishers are disregarding the rights of the artist.

I also had the idea that I would like to change such attitudes and so responded to their inquiries and requests in detail, but in the end I often had to change this editing policy. Also, because of a mistake or misunderstanding on my part, I ended up upsetting a couple of people, to whom I sincerely apologize. I may have upset more people out there though they are unbeknown to me. I had hoped to do my best and regret that this happened.

I started off thinking of the banners as pieces of art but I came to realize that behind them are the creators. I felt the great burden of responsibiity of having to respond to the wishes of almost 1,000 people.

By the time I knew how to improve the book and realized that my original methodology had been overoptimistic, I was at the point of no return. Although I had wanted to start from the beginning all over again, instead I had to compromise to a certain extent because I had to promised to make the publishing deadline.

For those who allowed me to publish their work, they may feel some dissatisfaction with the book when they see how it turned out. If this is the case, I hope that they will tell me. I will endeavor to incorporate their ideas in the next book.

Even though compromises were made, I think that the publishing of the book is an achievement in itself. And I feel that at least it is a book that people will enjoy looking at and be glad to have as I had hoped it would be.

We only managed to publish approximately 1,000 banners in the end because we ran out of time. This represents just a third of what is actually out there. Learning from the experiences I had in the making of this book, I hope that I can produce an upgraded second series in the future. I thank all the CG artists who supported this book for their cooperation and words of encouragement.

Lastly, I would like to thank Graphic-sha Publishing Co., Ltd. and Mr. Okuda for giving me the chance to realize my ideas into the book, Ponkichi for working up to the last minute on the layout, Ms. Kristin Bradley and Ms. Atsuko Nozawa for their translation. And now, Miwa-chan, I thank you for standing by me and may you always do so.

いとうあき=クイッカマン=あきびょん
Aki Ito = QuickcaMan = akibyon
http://www.ra.sakura.ne.jp/~akibyon/

デザイン+レイアウト:
本吉豊徳=panju=ぽんきち
Design and layout:
Toyonori Motoyoshi = Ponkichi

英訳:
クリステン・ブラッドリー、野澤敦子
English translation:
Kristin Bradley and Atsuko Nozawa

ISBN4-7661-1066-8

Graphic-sha Publishing Co., Ltd.
1-9-12 Kudan-kita Chiyoda-ku Tokyo 102-0073 Japan
Telephone 03-3263-4318 Fax 03-3263-5297

美少女CG網站

定價：450元

出版者：新形象出版事業有限公司
負責人：陳偉賢
地　址：台北縣中和市中和路322號8Ｆ之1
門　市：北星圖書事業股份有限公司
永和市中正路498號
電　話：29229000（代表）　ＦＡＸ：29229041

原　著：伊東秋
編譯者：新形象出版公司編輯部
發行人：顏義勇
總策劃：范一豪
文字編輯：賴國平・陳旭虎
封面編輯：許得輝

總代理：北星圖書事業股份有限公司
地　址：台北縣永和市中正路462號5F
電　話：29229000（代表）　ＦＡＸ：29229041
郵　撥：0544500-7北星圖書帳戶
印刷所：皇甫彩藝印刷股份有限公司

行政院新聞局出版事業登記證／局版台業字第3928號
經濟部公司執／76建三辛字第21473號

西元2000年2月　第一版第一刷

國家圖書館出版品預行編目資料

美少女CG網站=Anime CG Web banner collection
／伊東秋原著；新形象出版公司編輯部編譯．
--第一版．--臺北縣中和市：
新形象，1999〔民88〕
面：　公分

ISBN 957-9679-66-5（平裝）

1. 網際網路--站臺

312.91653　88013827